Problem-Solving in Conservation Biology and Wildlife Management

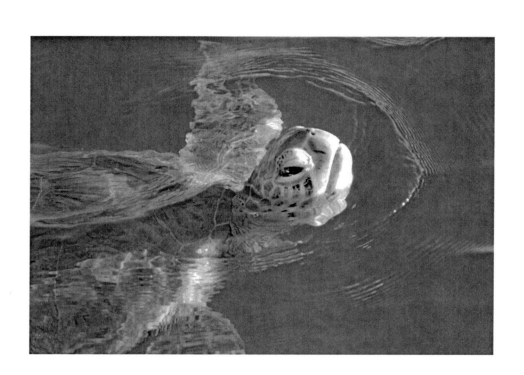

Problem-Solving in Conservation Biology and Wildlife Management

Exercises for Class, Field, and Laboratory

James P. Gibbs
Malcolm L. Hunter, Jr.
Eleanor J. Sterling

Blackwell
Publishing

© 2008 by James P. Gibbs, Malcolm L. Hunter, Jr., and Eleanor J. Sterling
© 1998 by Blackwell Publishing

BLACKWELL PUBLISHING
350 Main Street, Malden, MA 02148-5020, USA
9600 Garsington Road, Oxford OX4 2DQ, UK
550 Swanston Street, Carlton, Victoria 3053, Australia

First edition published 1998 by Blackwell Publishing
Second edition published 2008

2   2009

Library of Congress Cataloging-in-Publication Data

Gibbs, James P.
Problem-solving in conservation biology : exercises for class, field, and laboratory / James P. Gibbs, Malcolm L. Hunter, Jr., Eleanor J. Sterling. – 2nd ed.
    p. cm.
  Includes bibliographical references and index.
  ISBN 978-1-4051-5287-7 (pbk. : alk. paper) 1. Conservation biology. 2. Wildlife management.
3. Conservation biology–Problems, exercises, etc. 4. Wildlife management–Problems, exercises, etc.
I. Hunter, Malcolm L. II. Sterling, Eleanor J. III. Title.
  QH75.G53 2008
  577–dc22

                                    2007016279

A catalogue record for this title is available from the British Library.

Set in 10.5/13pt Janson
by SPi Publisher Services, Pondicherry, India

For further information on
Blackwell Publishing, visit our website:
www.blackwellpublishing.com

Contents

Preface

If you are a student today you have an opportunity to play a significant role in how the biodiversity crisis plays out. Although the long-term trajectory of the human population is unclear, during the next few decades the level at which it might eventually stabilize will become increasingly clear. Of course, enormous expanses of the environment have been fundamentally transformed by our activities, yet much is left that merits conserving. How we conserve that which remains, and how we choose to live upon and in some cases restore the rest, will determine the fates of millions of wild species. Much of this will transpire during your career.

While you can learn a good deal from attending class lectures, what will be most useful to you are practical experiences. This is because conservation biology and wildlife management are, more than anything, about the application of ideas to the solving of problems. One can go to great effort to discuss all the dimensions of the biodiversity crisis, carefully enumerate the individuals in an endangered population, or methodically poll the public on its attitudes toward wild life. Ultimately what matters most is putting this information into action. This is the biggest challenge that any practicing conservationist faces.

We have generated this book expressly for the students and teachers of conservation biology and wildlife management who want to have an impact beyond the classroom. The book originated from our collective sense that "learning by doing" is the most effective, fun, and durable way to develop into a professional. A so-called problem-based learning approach worked best for us when we were students. Now we wish to share this engaging learning approach with you.

We have created a set of exercises that addresses problems spanning a wide range of conservation issues: genetic analysis, population biology, ecosystem management, the public policy process, and more. Some can be used as simple homework exercises for individuals working alone. Others are lengthy, group exercises. All carry a message about "making it happen," that is, how to take what you have learned in an exercise and have an impact in the larger world.

The first edition of this book was published in 1998. In the interim, the book has been purchased and used by many around the world, enough to warrant a second edition. We now have more experience in developing exercises that "work." Approximately two-thirds of the material in this second edition is new or dramatically revised from the first.

Our target audience is upper-level college undergraduates, early-stage graduate students, and possibly some practicing professionals. While the book might best complement an existing conservation biology or wildlife management lecture course, it can contribute to a variety of courses, and has, for example, been adopted for a re-training course for secondary school teachers and a field-based natural history course.

We view conservation biology and wildlife management as complementary fields, and have therefore included exercises applicable to both. The two fields contrast mostly in terms of emphasis. Conservation biology views all of nature's diversity as important and having inherent value, whereas traditional wildlife management operates from a somewhat more utilitarian perspective with a primary objective of providing recreational resources for people, including sustained yields of harvested species, especially birds and mammals. Both fields recognize the need to integrate the contributions of non-biologists (economists, sociologists, political scientists) to conserve wild species. This commonality distinguishes both fields from the pure sciences. Because of the blurred distinctions between these fields we have intentionally not tried to identify which exercises are more suitable to a conservation biology class versus a wildlife management class.

The book has been designed to accompany any of the main-stream conservation biology and wildlife management texts. Instructors should be aware that they need a copy of the accompanying instructors' manual to make certain exercises succeed. To secure a copy, see "Important note about the instructor's manual" below.

Copying

We are well aware that the cost of textbooks often leads both students and faculty to copy portions of textbooks illegally. The temptation to do this is particularly great with lab texts in which not all of the exercises will be used. We have tried to minimize this temptation by keeping the price of the book low. We are also aware that many scanned, electronic copies of exercises from the first edition of the book are posted on course websites on the internet. This is flattering yet frustrating as it undercuts not only our efforts but also any publisher's interest in books such as this. Perhaps it will help some people to avoid the temptation of photocopying to know that all of the royalties from this book have been dedicated to conservation: two fellowship funds for natural resource and biology students from developing countries.

Important Note About the Instructor's Manual

Instructors should be aware that they need a copy of an accompanying instructor's manual to make many exercises succeed. While developing this book we compiled a companion electronic document with the answers to all the exercises as well as many tips and suggestions. We wish to manage the distribution of this manual to instructors of classes. To receive an electronic copy, please send an email message to James Gibbs at jpgibbs@esf.edu indicating:

- your institution and position
- the course name and number, and
- approximate number of students in the class.

With this information we will arrange a web download or email transmission of the manual.

As we regard this as an evolving project please also send along suggestions and criticisms. We would especially like to hear about ways to improve these exercises. We know that many teachers of conservation biology and wildlife management courses have put together similar exercises for their classes. For possible inclusion in a future edition of the book, please send them along to us.

The Book's Website

This book has an accompanying website with the data sets and other resources to support many of the exercises presented herein: www.blackwellpublishing.com/gibbs

Acknowledgments

We are most grateful to the students who worked through earlier, less polished versions of these exercises. Most of the exercises have been "field-tested" and greatly improved by students at the State University of New York's College of Environmental Science and Forestry, Columbia University, the University of Maine, and Yale University. Also, Erin McCreless and Brian Weeks kindly assisted with exercise preparation.

A number of people have either authored Exercises or written them with us. We are extremely pleased to include their excellent contributions here:

Margret C. Domroese, American Museum of Natural History, New York, USA
Jennifer Griffiths, American Museum of Natural History, New York, USA
Luigi Guarino, Secretariat of the Pacific Community, Suva, Fiji
Ian J. Harrison, American Museum of Natural History, New York, USA
Robert J. Hijmans, International Rice Research Institute, Los Baños, Philippines
Andy Jarvis, Bioversity International and International Centre for Tropical Agriculture (CIAT), Cali, Colombia
Thane Joyal, Law Offices of Joseph J. Heath, Syracuse, New York, USA
Michael E. Meredith, Wildlife Conservation Society, Kuching, Malaysia
Viorel Popescu, State University of New York College of Environmental Science and Forestry, Syracuse, New York, USA
Richard B. Primack and Brian Drayton, Boston University, Boston, USA
Pablo Ramirez de Arellano, Bioforest S.A., Concepción, Chile
Robert S. Seymour, University of Maine, Orono, Maine, USA
Krishnan Sudharsan, Michigan State University, East Lansing, Michigan, USA

We would also like to highlight the direct and indirect contributions to this effort by the staff of the Network of Conservation Educators and Practicioners (NCEP), particularly Nora Bynum and Ian Harrison of the American Museum of Natural History. NCEP is a global project to improve the practice of biodiversity conservation by improving training in biodiversity conservation. Some of the exercises herein are adapted directly from NCEP materials and you are encouraged to consult the NCEP website to access many related materials, including class presentation materials, topic syntheses, and even more exercises: http://ncep.amnh.org/

Drawings were produced by Debbie Maizels and the staff of Emantras.

Last, but certainly not least, at Blackwell Publishing Rosie Hayden patiently guided us through an extended publication process and Janey Fisher did an extraordinary job scrutinizing the manuscript during the copy-editing process for this 2nd edition.

Images at chapter and part openings: credits

Front (p ii): Green turtle at Monkey Mia, Western Australia. Image: Malcolm L. Hunter, Jr.

Introduction: Kihansi Gorge, Udzungwa Mountains, Eastern Arc, Tanzania. Image: James P. Gibbs.
  1 *What is Biodiversity?* Araneidae. Image: Berland (1955).
  2 *What is Conservation Biology?* Mountain ash (a species of *Eucalyptus*), Victoria, Australia. Image: Malcolm L. Hunter, Jr.
  3 *Why is Biodiversity Important?* Giant anteater, Roraima, Brazil. Image: James P. Gibbs.

Genes: PCR amplicons for representatives of 94 fish families. Image: Ivanova, N. V., *et al*. Molecular Ecology Notes 7:544–548.
  4 *Population Genetics:* Orchid, unknown spp., Roraima, Brazil. Image: J. P. Gibbs.
  5 *Genetic Drift:* Eastern barred bandicoot. Image: Hunter and Gibbs (2007).
  6 *Pedigree Management:* Triple nose-leaf bat. Image: Blanford (1888).
  7 *Landscape Genetics:* American toad, New York State, USA. Image: James P. Gibbs.

Populations: Leopard frogs, New York State, USA. Image: Nancy Karraker
  8 *Life Table Analysis:* Albatross, Antarctic Ocean. Image: Malcolm L. Hunter, Jr.
  9 *Population Viability Analysis:* Galapagos Penguins, Galapagos Islands, Ecuador. Image: James P. Gibbs.
  10 *Habitat Loss and Fragmentation:* White trillium, New York State, USA. Image: Donald J. Leopold.
  11 *Diagnosing Declining Populations: Opuntia* spp. Image: James P. Gibbs.
  12 *Estimating Population Size with DISTANCE:* Orang utan, Sepilok Rehabilitation Center, Sabah. Image: Michael E. Meredith.
  13 *Analyzing Camera Trap Data:* Golden cat. Image: Zoo Heidelberg/Ales Toman.
  14 *Estimating Population Size with MARK:* Karanth and Nichols (2002). © Center for Wildlife Studies.

Species: Atlantic puffin. Image: Malcolm L. Hunter, Jr.
  15 *Estimating "Biodiversity":* Various Coleoptera. Image: W. S. Blatchley, The Nature Publishing Company, Indianapolis, Indiana (1910).
  16 *Designing a Zoo:* Candidate species. Image: Hunter and Gibbs (2007).
  17 *Plant Reintroductions:* Blue lupine ready for outplanting. Image: James P. Gibbs.
  18 *Edge Effects:* Artificial nest with hen's egg. Image: James P. Gibbs.

Ecosystems and Landscapes: Acadia National Park, Maine, USA. Image: Malcolm L. Hunter.

19 *Ecosystem Fragmentation:* Fragmentation stage. Image: Hunter and Gibbs (2007).

20 *Forest Harvesting:* Macaw, Roraima, Brazil. Image: James P. Gibbs.

21 *Protected Areas:* Working landscape. Image: Hunter and Gibbs (2007).

22 *Island Biogeography:* Species-area relationships. Image: Hunter and Gibbs (2007).

23 *GIS for Conservation:* Wild potato (*Solanum megistacrolobum* subsp. *toralapanum*), Toralapa, Cochabamba, Bolivia. Image: Robert Hijmans.

24 *Global Change:* Mink frog, Adirondack Mountains, New York, USA. Image: James P. Gibbs.

25 *Climate Envelope Modeling:* Wild peanut (*Arachis nitida*). Image: Karen Williams, U.S. Department of Agriculture.

Policy and Organization: Bottle-nosed dolphin, Monkey Mia, Western Australia. Image: Malcolm L. Hunter, Jr.

26 *Population, Consumption, or Governance:* Syracuse, New York, USA. Image: James P. Gibbs.

27 *Overconsumption:* SUNY-ESF, Syracuse, New York, USA. Image: James P. Gibbs.

28 *Conservation Values:* Opinion survey. Image: James P. Gibbs.

29 *Priority Setting:* Shavla River, Argut Nature Park, Altai, Russia. Image: James P. Gibbs.

30 *Commercial fishing in Galapagos National Park:* Pinzon Island, Galapagos National Park, Galapagos, Ecuador. Image: James P. Gibbs.

31 *Conservation Law:* Polar bears, Svalbard Islands. Image: Malcolm L. Hunter, Jr.

32 *Conservation Policy:* Formal letter. Image: James P. Gibbs.

PART 1

Introduction

CHAPTER 1

What is Biodiversity? Spiders as Exemplars of the Biodiversity Concept

James P. Gibbs, Ian J. Harrison, and Jennifer Griffiths

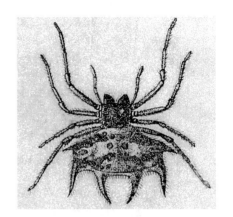

What is biodiversity? Ask anybody on the street if they have heard of the word "biodiversity" and you will usually get an affirmative answer. This is because the term has become very widely known – competing with some of the most popular terms from the media on the Internet. Popularly, the term resonates with the general public by conjuring up images of "tropical forest ringing with a cacophony of unseen frogs, insects, and birds; a coral reef seething with schools of myriad iridescent fishes; a vast tawny carpet of grass punctuated by herds of wildebeest and other antelope" (Hunter & Gibbs 2006). But then ask them to define it. Answers will range widely because the term is surprisingly misunderstood.

In practice "biodiversity" typically refers to the number and variety of distinct organisms (species) living on the Earth even though individual species are just one level of organization in living nature. Even with a focus simply on species diversity the concept of biodiversity and the practice of studying it can be overwhelming. The job of sorting out the many millions of species on Earth is the foundation of taxonomy, a subdiscipline within the larger discipline of systematics. The job of the taxonomist is to find an organized way for humans to make distinctions that parallel the distinctions that various organisms make themselves. Such distinctions are fundamental to conservation biology because if we are to maintain biodiversity we must understand how it is organized. In particular, we often need to identify which organisms will be affected by a particular conservation action. Moreover, we often need to focus on not just which species are different from one another but also consider *how* different they are.

Objective

- To explore the concept of biological diversity as it occurs at various taxonomic levels through the classification of life forms.

Procedures

Spiders are a highly species-rich group of invertebrates that exploit a wide variety of niches in virtually all the earth's biomes. Some species of spiders build elaborate webs that passively trap their prey whereas others are active predators that ambush or pursue their prey. Spiders represent useful indicators of environmental change and community level diversity because they are taxonomically diverse, with species inhabiting a variety of ecological niches, and they are easy to catch.

Sorting and Classifying a Spider Sample Collection

Spider collections have been assembled for you from 5 forest patches (sites 1 through 5) depicted on a map (Figure 1.1). Our first focus is on the spider sample collection for site 1 (Figure 1.2). The spiders were captured by a biologist traveling along transects through the forest patch, striking with a stick a random series of 100 tree branches. All spiders that were dislodged and fell onto an outstretched sheet were collected and preserved in alcohol. They have since been spread out on a tray for you to examine. The illustrations of the spiders collected are aligned in rows and columns so that it is easy to cut them out with scissors (if you wish) for subsequent examination, grouping, and identification.

The next task is to sort and identify the spiders. Look for external characters that all members of a particular group of spiders have in common but that are not shared by other groups of spiders. For example, look for characteristics such as leg length, hairiness, relative size of body segments, or abdomen patterning and abdomen shape. Describe briefly the set of characters unique to each group of morphologically indistinguishable spiders. These "operational taxonomic units" that you define will be considered separate species.

Assign each species a working name, preferably something descriptive. For example, you might call a particular species "spotted abdomen, very hairy" or

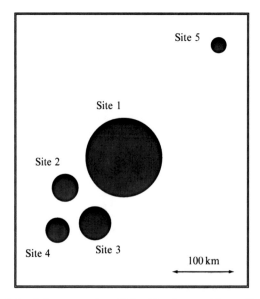

Fig. 1.1 Locations of the 5 forest patches ("sites") where spider collections were made.

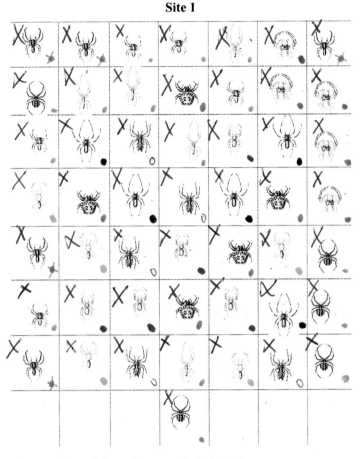

Fig. 1.2 Specimens collected from forest patch ("site") 1.

"short legs, spiky abdomen." Just remember that names that signify something unique about the species will be more useful to you. Construct a table listing each species, its distinguishing characteristics, the name you have applied to it, and the number of occurrences of the species in the collection.

Has Sampling at Site 1 been Adequate to Characterize the Spider Community?

Next ask whether this collection adequately represents the true diversity of spiders in the forest patch at the time of collection. Were most of the species present sampled or were some likely missed? This is always an important question to ask to ensure that the sample was adequate and hence legitimately characterizes the assemblage of spiders at a site.

You can address this question by performing a simple but informative analysis that is standard practice for conservation biologists who do biodiversity surveys. This analysis involves constructing a simple, so-called "collector's curve" (Colwell & Coddington 1994). These plot the cumulative number of species observed (y-axis) against the cumulative number of individuals classified (x-axis). The collector's curve is an increasing function with a slope that will decrease as more individuals are classified and as fewer species remain to be identified (Figure 1.3). If sampling stops while the collector's curve is still rapidly increasing, sampling is incomplete and many species likely remain undetected. Alternatively, if the slope of the collector's

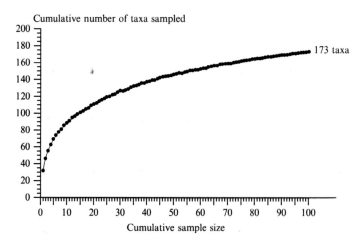

Fig. 1.3 An example of a collectors' curve. Cumulative sample size represents the number of individuals classified. The cumulative number of taxa sampled refers to the number of new species detected.

curve reaches zero (flattens out), sampling is likely more than adequate as few to no new species remain undetected.

To construct the collectors' curve for this spider collection, choose any specimen within the collection at random (if you cut up the spider collections with scissors put them in a hat or bag and shake them up). This will be your first data point, such that $x = 1$ and $y = 1$, because after examining the first individual you have also identified one new species. Next move in any direction (but consistently so) to a new specimen and record whether it is a member of a new species. In this next step, $x = 2$, but y may remain as 1 if the next individual is not of a new species or it may change to 2 if the individual represents a new species (different from specimen 1). Repeat this process until you have proceeded through all 50 specimens and construct the collector's curve from the data obtained (just plot y versus x). Does the curve flatten out? If so, after how many individual spiders have been collected? If not, is the curve still increasing? What can you conclude from the shape of your collector's curve for site 1 as to whether the sample is an adequate characterization of spider diversity at the site?

Contrasting Spider Diversity among Sites to Provide a Basis for Prioritizing Conservation Efforts

Now you are provided with spider collections from 4 other forest patches. The forest patches have resulted from fragmentation of a once much larger, continuous forest; a stylized map of the fragmented forest patches, showing their size and proximity to each other, is given in Figure 1.1. You will use the spider diversity information to prioritize efforts for the five different forest patches (including the data from the first patch which you have already classified). The additional spider collections are provided in Figures 1.4, 1.5, 1.6, and 1.7.

Again, tally how many individuals belonging to each species occur in each site's collection (use your classification of spiders completed for Site 1 during Level 1 of the exercise as a starting point with the caveat that you will "discover" some new species at the new sites). Data are most easily handled by constructing a table of species (rows) by site (columns). In the table's cells put the number of individuals of each species you found in the collection from each site. You can then analyze these data to generate different measures of community characteristics to help you to

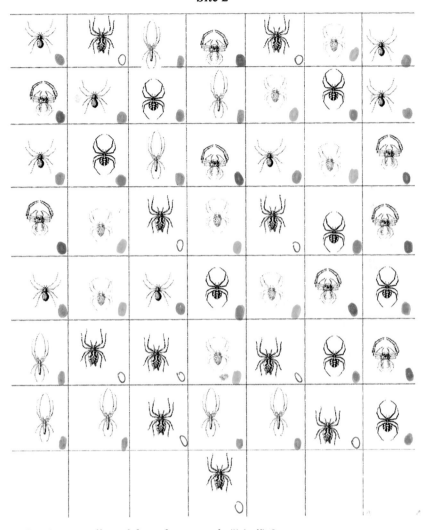

Fig. 1.4 Specimens collected from forest patch ("site") 2.

decide how to prioritize protection of the forest patches. Recall that you need to rank the patches in terms of where protection efforts should be applied, and you need to provide a rationale for your ranking.

Analyzing Community Diversity

You will find it most useful to base your decisions about how to prioritize these sites for protection on three community characteristics:

- species richness within each forest patch
- species diversity within each forest patch
- the similarity of spider communities between patches.

Species richness is simply the tally of different spider species that were collected in a forest patch. Species diversity is a more complex concept. We will use a standard index called Simpson Reciprocal Index, 1/D where D is calculated as follows:

$$D = \sum p_i^2$$

Site 3

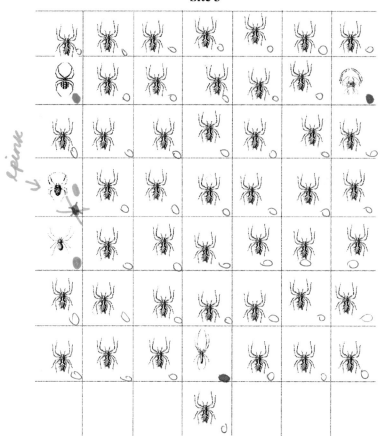

spink →

$=1/\left((1.5)^{2}+(.5)^{2}\right)=\dfrac{1}{0.5}$ or 2

Fig. 1.5 Specimens collected from forest patch ("site") 3.

where p_i = the fractional abundance of the *i*th species on an site. For example, if you had a sample of 2 species with 5 individuals each, Simpson Reciprocal Index = 1 / ((0.5)2 + (0.5)2) = 1/2 or 0.5. The higher the value, the greater the diversity. The maximum value is the number of species in the sample, which occurs when all species contain an equal number of individuals. Because this index not only reflects the number of species present but also the relative distribution of individuals among species within a community it can reflect how "balanced" communities are in terms of how individuals are distributed across species. Textbooks usually discuss this concept in terms of "evenness." As a result, two communities may have an identical complement of species, and hence species richness, but substantially different diversity measures if individuals in one community are skewed toward a few of the species whereas individuals are distributed more evenly in the other community.

Analyzing Community Distinctiveness

Diversity is one thing, distinctiveness is quite another. Thus another important perspective in ranking sites is how different the communities are from one another. We will use the simplest available measure of community similarity, that is, the Jaccard coefficient of community similarity (CC_J), to contrast community distinctiveness between all possible pairs of sites:

$$CC_J = c/S$$

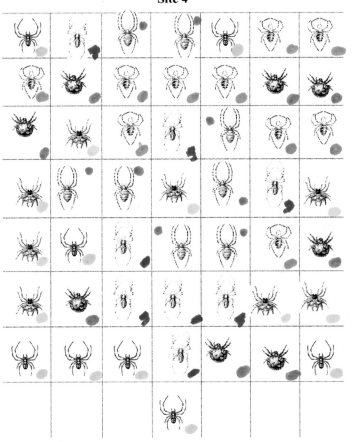

Fig. 1.6 Specimens collected from forest patch ("site") 4.

where c is the number of species common to both communities and S is the total number of species present in the two communities. For example, if one site contains only 2 species and the other site 2 species, one of which is held in common by both sites, the total number of species present is 3 and the number shared is 1, so $1/3 = 33\%$. This index ranges from 0 (when no species are found in common between communities) to 1 (when all species are found in both communities). Calculate this index to compare each pair of sites separately, that is, compare Site 1 with Site 2, Site 1 with Site 3, ..., Site 4 with Site 5 for a total of 10 comparisons. You might find it useful to determine the average similarity of one community to all the others, by averaging the CC_J values across each comparison in which a particular site is included.

Once you have made these calculations of diversity (species richness and Simpson's Reciprocal Index) and distinctiveness (CC_J), you can tackle the primary question of the exercise:

- How should you rank these sites for protection and why?
- Making an informed decision requires reconciling your analysis with concepts of biological diversity as it pertains to diversity and distinctiveness.

Your decisions can be based principally on your estimates of species richness, diversity, endemicity (species found at only one site), and community similarity. However, once you have made those decisions you might also want to look at the spatial arrangement of the forest patches and compare that to the species

Site 5

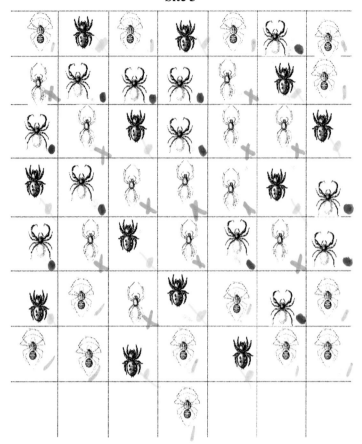

Fig. 1.7 Specimens collected from forest patch ("site") 5.

distributions. This might help you in your interpretation of the species distributions, and might give useful additional information for ranking the sites for protection.

Considering Evolutionary Distinctiveness

When contrasting patterns of species diversity and community distinctiveness, we typically treat each species as equally important, yet are they? What if a species-poor area actually is quite evolutionarily distinct from others? Similarly, what if your most species-rich site is comprised of a swarm of species that have only recently diverged from one another and are quite similar to species present at another site? These questions allude to issues of biological diversity at higher taxonomic levels. Only by looking at the underlying evolutionary relationships among species can we gain this additional perspective. We have provided below a phylogeny of the spider families that occur in your collections (a genuine phylogeny for these families based in large part on Coddington & Levi, 1991). In brief, the more closely related families (and species therein) are located on more proximal branches within the phylogeny.

First, you need to match your species to the ones shown for the different families in the tree of phylogenetic relationships (Figure 1.8). Do you see any patterns of distributions across the forest patches for related species (that is, species that belong to the same family)?

Next, look for any interesting patterns of distributions of the families across the forest patches. Based on the evolutionary relationships among these families, and

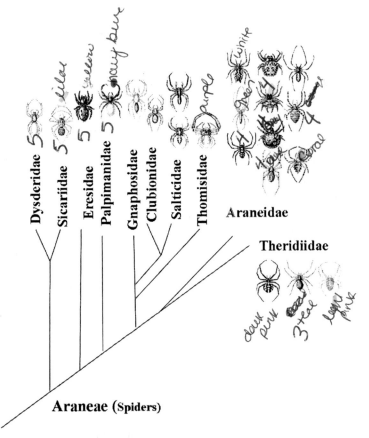

Fig. 1.8 Phylogenetic relationships among the spider taxa encountered on sites 1–5. From Coddington & Levi 1991.

their distribution in the forest patches, do you find any additional information that might help inform your decisions on prioritizing forest patches?

Expected Products

- A key to the species of spider identified (descriptive names and diagnostic characteristics, all ordered by family)
- A table of number of individuals tallied by species (rows) by site (columns)
- A table of the same format as the preceding one but with the proportion of total individuals at a site represented by each species at the site
- A table of the same format as the preceding one but with the proportions squared and then summed by site at the bottom of each site column. The reciprocal of the sums (the diversity measure) can then be easily calculated from these data
- A table (or matrix) of the average CC_J values of each site compared to the other four sites
- A "collectors curve" for site 1 and its interpretation in terms of the adequacy of sampling to characterize spider diversity at the site
- A summary (in a format outlined by your instructor) of key findings and a description of how you will go about prioritizing the sites for protection based on the information gathered, including a consideration of the phylogenetic perspectives that may or may not change your conclusion based solely on the community composition data.
- Responses in a form indicated by your instructor to the Discussion questions below.

Discussion

1 How would you rank the sites for protection and why?

2 Why would some conservation biologists make decisions emphasizing species endemism over species diversity?

3 Why does it matter if there are more species in one area than in another?

4 For the most part, conservation decisions as fundamental as where to put a reserve or which areas to restore should not be made on the basis of one taxon. What are the advantages and disadvantages of using an indicator taxon such as spiders as a proxy for comprehensive biodiversity surveys?

5 This exercise focuses on fully recognized species. If the site that you prioritized last for protection had a wide variety of subspecies present for each of the species it hosted, whereas on all the other islands each species was represented by a single taxon, would you change your recommendations?

6 What other site-level information would you find useful to assist the prioritization process beyond that related to species present and their evolutionary relationships?

Making It Happen

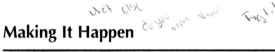

The work you have just completed is not simply an academic exercise. It used approaches very similar to those used currently all over the world to assess site-level biological diversity as a basis for making decisions about which areas to prioritize for protection. For example, investigate Conservation International's Rapid Assessment Program (RAP) website (information below) and scan through some of the trip reports. You will see many similarities to what you did here. You might find an opportunity one day to participate in such efforts. In the meantime, just look around your region — you will undoubtedly find conservation groups that need this kind of information for planning purposes. Moreover, there are likely other groups that devote a few days a month to going out in the field and observing and identifying various taxa – joining these groups is a fine way to build your skills as taxonomist. Bird watching is the most common activity, but mushroom forays, wildflower walks, and amphibian and reptile explorations also are often organized. Experienced members of these groups know that to persist they must recruit new members, and are often eager to instruct young people on the "ins" and "outs" of identifying a certain group of organisms. Because many amateur taxonomists are also avid conservationists, participating in these groups also permits one to make important contacts within the conservation community.

Further Resources

Some useful discussions of the importance of phylogenetic analyses in biodiversity conservation are provided by Stiassny (1992), Stiassny & de Pinna (1994), and Benstead et al. (2003). For more on the issue of taxonomy and systematics in the biodiversity crisis, see Vane-Wright et al. (1991). To see real-world examples of what you just did, consult Conservation International's Rapid Assessment Program (RAP) website: www.conservation.org/xp/CIWEB/programs/rap/

What is Conservation Biology? An Analysis of the Critical Ecosystem Partnership Fund's Strategies and Funding Priorities

James P. Gibbs

Attend a meeting of the Society for Conservation Biology and you are likely to meet economists and ecologists, geneticists and geographers, philosophers, foresters, and fisheries managers. People can wear many different hats and still say, "I am a conservation biologist." Conservation biology is a remarkably diverse discipline that sits squarely at the intersection of many other fields. Of course, one cannot define a discipline solely in terms of other intersecting disciplines. Conservation biology is best defined by its overarching goal, maintaining the planet's biological diversity. This focus creates a unique niche for conservation biologists, nestled within the larger arena of environmental management, applied ecology, and natural resources management. More and more people are finding a home in that niche as we become increasingly sensitive to the plight facing most of the earth's species and ecosystems.

But just what is Conservation Biology? Conservation biology is a crisis mission focused on saving life on earth. Despite this noble cause, the discipline still fails to "ring a bell" with much of the general public. One of the reasons is that conservation biology is indeed a mish-mash of many disciplines. Science in general and biology in particular play a big role in conservation biology yet the field extends into many other disciplines. These include finance, law, sociology, organization management, communications, and education; in other words, the "human dimensions" of conservation biology (Jacobson 1990, Soulé 1985). Expertise in these latter fields is what gives conservation biologists traction in the real world. It's been often said that conservation biology is as much about changing people's habits as it is about saving nature.

So what is the precise mix of disciplines? What do conservation biologists actually do? If you are preparing for a career in conservation biology, what skills should you develop? A schematic model developed by Susan Jacobson (1990) has frequently been used to depict the interacting fields that constitute conservation biology (Figure 2.1),

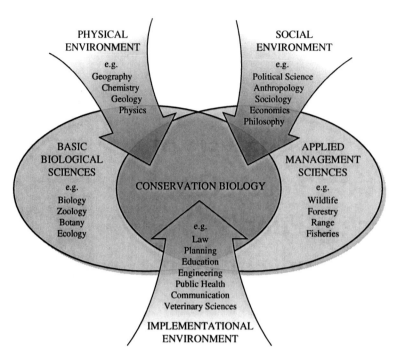

Fig. 2.1 Schematic depicting the interaction of disciplines that together represent the field of conservation biology (redrawn after Jacobson 1990).

and gives us a starting point for analyzing just what conservation biologists do. This model indicates that students seeking a career in conservation biology need to develop an unusually broad outlook, marrying a focus on basic biological sciences and its application via the natural resources to a human-centered focus on economics, politics, law, and communication, which together represent the political arena in which all conservation efforts must operate. Conceptual models are useful but perhaps most useful is breaking down what conservation biologists are doing right now to stem the loss of biological diversity. In this exercise you will evaluate current strategies used by one of the largest and most ambitious conservation programs operating around the world. In analyzing the strategic emphases of this program, we hope to provide you with a timely view of what conservation biology is all about.

Objectives

- To learn what conservation biologists do
- To generate an appreciation of the complexities that underlie most conservation issues
- To identify the diverse skills required for a career in conservation.

Procedures

In this exercise we focus on the Critical Ecosystem Partnership Fund (CEPF). The CEPF advances biodiversity conservation at the global scale. Its main goal is to catalyze interactions among diverse groups working in the conservation and thereby develop a comprehensive, coordinated approach to conservation problems. Because the CEPF seeks to achieve the highest returns on conservation investments it focuses primarily on regions that are considered biodiversity "hotspots." These are the 25 or so regions that cover just 1.4 percent of the Earth's land surface yet host more than 60 percent of the planet's terrestrial species diversity. To be eligible for funds projects must be (i) within a biodiversity hotspot, (ii) within a developing country that has ratified the Convention on Biological Diversity, and (3) undertaken by applicants that represent nongovernmental organizations, community groups, or private-sector partners.

CEPF is a joint initiative of Conservation International, the Global Environment Facility, the Government of Japan, the John D. and Catherine T. MacArthur Foundation, and the World Bank, with some $125 million committed to date. Funds are administered by the non-profit conservation group Conservation International. Funds are provided in the form of grants to individual recipient organizations consistent with overall CEPF strategy. An immense amount of thought has gone into setting priorities and strategy to ensure that the many millions of dollars allocated to the effort produce tangible benefits for biodiversity and the local people associated with it. Because the CEPF represents a direct manifestation of the field of conservation biology it can be illuminating to focus on what strategies it has devised.

As a tool to organize your thinking, Jacobson's (1990) schemata shows graphically the relationship between conservation biology and other disciplines (Figure 2.1). To see how this model plays out in reality we will use it as the starting basis for an analysis of the CEPF strategies. The procedure is straightforward. We will crosswalk Jacobson's figure (and any variations of it that you can come up with) with the conservation directions and investment priorities established by the CEPF for five recognized global biodiversity hotspots: Eastern Arc Mountains and Coastal Forests of Tanzania and Kenya (Box 2.1), Cape Floristic Region (Box 2.2), Caucasus (Box 2.3), Southern Mesoamerica (Box 2.4), and the Mountains of Southwest China (Box 2.5).

Box 2.1 CEPF strategic funding directions and investment priorities in the Eastern Arc Mountains and Coastal Forests of Tanzania and Kenya (2003–2008)

1. Evaluate community-based forest management initiatives in the hotspot to determine best practices.
2. Promote nature-based, sustainable businesses that benefit local populations in the hotspot.
3. Explore possibilities for direct payments and easements (Conservation Concessions) for biodiversity conservation in the hotspot and support where appropriate.
4. Build the capacity of community-based organizations in the hotspot for advocacy in support of biodiversity conservation at all levels.
5. Support cultural practices that benefit biodiversity in the hotspot.
6. Research and promote eco-agricultural options for the local populations of the hotspot.

(Continued)

Box 2.1 (*Continued*)

7. Assess potential sites in the hotspot for connectivity interventions.
8. Support initiatives that maintain or restore connectivity in the hotspot.
9. Monitor and evaluate initiatives that maintain or restore connectivity in the hotspot.
10. Support best practices for restoring connectivity in ways that also benefit people.
11. Refine and implement a standardized monitoring program across the 160 eligible sites.
12. Support research in the less studied of the 160 eligible sites in the hotspot.
13. Monitor populations of Critically Endangered and Endangered Species in the hotspot.
14. Support research in the hotspot to facilitate Red List assessments and re-assessments for plants, reptiles, invertebrates and other taxa.
15. Compile and document indigenous knowledge on hotspot sites and species.
16. Support awareness programs that increase public knowledge of biodiversity values of the hotspot.
17. Support targeted efforts to increase connectivity of biologically important habitat patches.
18. Support efforts to increase biological knowledge of the sites and to conserve critically endangered species.
19. Establish a professional resource mobilization unit, within an appropriate local partner institution, for raising long-term funds and resources for the hotspot.
20. Utilize high-level corporate contacts to secure funding from the private sector for the hotspot.
21. Train local NGOs and community-based organizations in fundraising and proposal writing.

Box 2.2   CEPF strategic funding directions and investment priorities in the Cape Floristic Region (2003–2008)

1. Identify and design innovative mechanisms and strategies for conservation of private, corporate, or communal landholdings within biodiversity corridors.
2. Support private sector and local community participation in the development and implementation of management plans for biodiversity corridors.
3. Especially within the Gouritz and Cederberg corridors, identify priority landholdings requiring immediate conservation action.
4. Promote civil society efforts to establish and support biodiversity-based businesses among disadvantaged groups, in particular in areas surrounding the Gouritz and Baviaanskloof corridors.
5. Implement best practices within industries affecting biodiversity in the CFR, e.g. the wine and flower industries.

(Continued)

Box 2.2 (*Continued*)

6. Support civil society efforts to consolidate data to support appropriate land use and policy decisions.
7. Integrate biodiversity concerns into policy and local government procedures in priority municipalities.
8. Improve coordination among institutions involved in conservation of CFR biodiversity corridors through targeted civil society interventions.
9. Support internships and training programs to raise capacity for conservation, particularly targeting previously disadvantaged groups.
10. Support initiatives to increase technical capacity of organizations involved in CFR conservation, particularly in relation to the priority geographic areas.

Box 2.3   CEPF strategic funding directions and investment priorities in the Caucasus (2003–2008)

1. Promote transboundary cooperation by carrying out joint initiatives and harmonizing existing projects to conserve border ecosystems and species and site outcomes.
2. Support existing efforts to create new protected areas and wildlife corridors through planning processes and co-financing efforts.
3. Develop and implement management plans for model protected areas with broad participation of stakeholders.
4. Provide funding for research and implementation of the Caucasus Red List re-assessments, particularly for poorly represented taxas such as plants, invertebrates, reptiles, and fish.
5. Focus small grant efforts on supporting efforts to conserve 50 globally threatened species in the hotspot.
6. Provide support to conservation agencies specifically to improve implementation of international conventions such as the Convention on Biological Diversity, the Convention on International Trade in Endangered Species and the Ramsar Convention on Wetlands.
7. Evaluate and implement models for sustainable forestry, water use and range management.
8. Focus small grant efforts on supporting existing NGOs to undertake projects focused on developing alternative livelihoods, such as ecotourism, collection of non-timber forest products and sustainable hunting and fishing.
9. Support civil society efforts to mitigate, participate in, and monitor development projects.
10. Develop local capacity to train environmental journalists and develop incentives to write on environmental issues, targeting decisionmakers in particular.
11. Develop a communications campaign to increase environmental awareness.

Box 2.4   CEPF strategic funding directions and investment priorities in Southern Mesoamerica (2003–2008)

1. Create a coordinating group, led by the NGO community, that will guide conservation actions in the Cerro Silva-Indio Maiz-La Selva Corridor.
2. Support NGO efforts to evaluate modalities for establishing additional private conservation areas to integrate connectivity among key areas.
3. Support civil society efforts and community efforts to establish best practices in coffee, cocoa, and tourism in areas of potential connectivity.
4. Implement awareness programs focused on flagship species in order to improve public understanding of the value of biodiversity.
5. Establish an emergency fund to support projects that will help protect critically endangered species.
6. Create participatory management plans in target areas and provide opportunities for civil society to participate in government led planning processes.
7. Establish the Maquenque National Park in northern Costa Rica.
8. Support civil society efforts to establish protected areas within the Ngobe-Bugle indigenous territory.
9. Support efforts by the NGO and private sector community to provide financial incentives for private reserves and conservation set-asides.
10. Support targeted civil society efforts to implement discreet elements of existing management plans.

Box 2.5   CEPF strategic funding directions and investment priorities in the Mountains of Southwest China (2003–2008)

1. Define 5- and 10-year map-based conservation outcomes for the hotspot through a collaborative, participatory approach.
2. Support projects that utilize scientific tools to evaluate changes in land cover, spatial relationships, and ecosystem health.
3. Establish a mechanism to monitor and evaluate the effectiveness of the site-specific projects and ensure adaptive management and sharing of lessons learned.
4. Provide resources to track human-induced environmental trends and high-resolution monitoring to report on site-specific impacts.
5. Conduct scientific research and socioeconomic analysis to better understand biodiversity and conservation issues and threats in the region.
6. Improve the credibility and scientific methodology used for biodiversity conservation research in this hotspot.
7. Enact effective nature reserve and community resource management.
8. Develop ecotourism and environmental education as a tool to support biodiversity conservation.
9. Undertake ecosystem restoration, especially filling in the gaps in existing governmental programs.
10. Projects to reduce illegal and other unsustainable wild animals and plants trade.

(Continued)

Box 2.5 (*Continued*)

11. Promoting biodiversity-friendly "green" production or harvest of traditional Chinese medicines.
12. Assess, develop and implement a series of training programs based on the training needs in the region. Training could focus on a number of topics including reserve management, the fundamentals of green businesses, business management for conservation and environmental education.
13. Provide resources for individuals in the region to participate in training opportunities.
14. Establish a trainers' training program in the region to multiply transfer of skills and knowledge to conservation professionals in the region.
15. Demonstrate best-case innovative approaches for integrating biodiversity concerns into local, regional and national development programs.
16. Collect and disseminate information about biodiversity and socioeconomic benefits of conservation to improve implementation of existing government initiatives and influence national policies.
17. Communicate successful examples of innovative approaches to public-private efforts to better integrate biodiversity conservation into governmental efforts.
18. Provide funding to individuals and institutions for research analysis or small-scale activities that will help build the conservation capacity of civil society and/or yield measurable mitigation of threats.
19. Provide technical support to trainees to enable better design and implementation of small on-the-ground projects.

These are merely examples drawn from biodiversity hotspots around the world to give you a sense of the complexities involved in actually implementing conservation biology. Inspecting these priorities will enable you to elucidate just which topics among the many that fall under the umbrella of conservation biology are most germane to its practice. Review all available information about the strategies for the selected hotspots with a copy of the Jacobson figure at hand and make a list of all the topics that are relevant to each hotspot. Tally these in a summary table to examine the overall distribution of priorities among topic areas.

For example, your table might array the hotspots (Eastern Arc Mountains and Coastal Forests of Tanzania and Kenya, Cape Floristic Region, Caucasus, Southern Mesoamerica, and Mountains of Southwest China) as column headers and strategic areas as rows. You should devise your own topic areas based on Figure 2.1 as a starting point but here are some suggestions: Biodiversity research, assessment, or management, Communications, Community-based activities, Conservation finance, Economic enterprise, Information management, Inventory and monitoring, Organizations and policy, and Protected areas and corridors. Note that some hotspot-specific strategies can have multiple foci and as such are best tallied more than once; for example, a project to boost sustainable livelihoods of farmers in potential biological corridor zones might best be tallied as both economic enterprise and protect areas/corridors.

Expected Products

- A tabular summary of the focal areas for biodiversity conservation as highlighted by the CEPF for each hotspot and an accompanying synthesis (in a format indicated by your instructor) of what professional skills and training would seem most useful to be able to participate effectively in these efforts.
- Responses in a form indicated by your instructor to the Discussion questions below.

Discussion

1 What are the most heavily represented topic areas in the CEPF strategies? What are the least? Are you surprised by the results?
2 Why do you think topic areas differ so much from one hotspot to another? What are the social, economic and cultural explanations for this variation?
3 In a conservation context, how might the relative importance of different disciplines changed over the last 50 years? How are they likely to change in the future?
4 Is the CEPF and the activities it supports an adequate representation of what conservation biology is all about? Why or why not?

Making It Happen

This exercise is intended to let you see what a "real-world" biodiversity conservation program looks like. If you seek to get involved in any conservation problem, be it local or global, you will need a thorough understanding of an intricate web of issues; an analysis such as this is a logical place to begin. At this point in your career you might also use your analysis as the basis for examining the curriculum at your school . . . are you developing the skills you need?

Further Resources

For further reading about the definition of conservation biology see any of the major conservation biology textbooks, including Caughley and Gunn (1996), Cox (2005), Groom et al. (2006), Hunter and Gibbs (2006), Primack (2004a and 2004b), Pullin (2002), Sutherland (2000), and Van Dyke (2002). One of the best ways to understand what falls under the umbrella of conservation biology is to read the two key journals, *Conservation Biology* and *Biological Conservation*. The primary professional organization is the Society for Conservation Biology; the URL for their website is www.scb.org. For more information on the CEPF consult its website: http://www.cepf.net/xp/cepf/.

CHAPTER 3

Why is Biodiversity Important? Why Is It Threatened? An Exploration with the IUCN "Red List" of Threatened Species

James P. Gibbs

Although species extinctions are part of the evolutionary process, current extinction rates are much greater than at any time in the last several million years. To address this issue, we need to understand the major threats to biodiversity. Many human activities threaten biodiversity either directly or indirectly, and virtually all current extinctions are due to human activities. However, each species faces its own specific suite of threats. Moreover, species in different regions of the world are more prone to some threats than others. If we are to mitigate these threats we must first understand them.

We also need to ask: What is the importance of biodiversity? These two questions are part and parcel of the same issue...if biodiversity's importance cannot be articulated, then it is hard to convince others to reduce threats to it. Fortunately, there are ways to try to bring the issue of why biodiversity is important into better focus. As in human life, sometimes how we value others emerges best when we are about to lose them. A friend moves to a new city or a grandparent dies. Their passage often provokes reflection upon what they meant to us and the ways in which they were important to us. Similarly, many wild species are about to depart from our existence, and their passage can force us to come to better grips with whether their extinction and hence their existence has value to us or not.

The World Conservation Union maintains a list of imperiled and extinct species that can serve as a useful point of departure for an examination of both threats and values. The list, known as the Red List of Threatened Species, or "Red List" for short, is continually being compiled and updated for species all over the world. It is easy to get a list of all the species known to be currently threatened with extinction in a country of interest to you. With this list in hand, you can become familiar with these species and speculate how their fate is important to us. We can also then contrast the causes of species imperilment in a particular country of interest with those in the world at large. From this assessment, you can develop an understanding of the primary threats to biodiversity both locally and globally, as well as what potential values associated with these species are at stake.

Objectives

- To explore the complex ways that imperiled wild species have "value"
- To understand regional versus global threats to biodiversity
- To become familiar with the IUCN Red List of threatened species as a resource for understanding species' changing status around the world.

Procedures

Securing the List of Imperiled Species

The Red List is constantly being updated and modified. For the most recent version, access the internet and go to: www.iucnredlist.org/ and select "Search" (you can choose Expert Search but for the purposes of this exercise a simple search will suffice).

Values Assessment

On the Search screen, first select the Red List categories of interest. Select CR (critically endangered: facing an extremely high risk of extinction in the wild), EN (endangered: facing a very high risk of extinction in the wild) and VU (vulnerable: facing a high risk of extinction in the wild). If you are having trouble selecting multiple categories, see the "help" button by the Red List Categories. These three categories include the most threatened, extant species among all those with sufficient data available to categorize them (Figure 3.1). You are strongly encouraged to become familiar with the different IUCN list categories and criteria at: www.iucn-redlist.org/info/categories_criteria2001. The reason is that IUCN list categories and criteria are part of the common language of most conservation biologists.

Under "Select one or more countries" choose any country of interest to you. This might be your home country or another that simply intrigues you. Now choose Search and you will generate a list of species. "Sort" them by Kingdom and then inspect the list. These are the species that, without concerted conservation effort, are soon to be lost. You can learn what little may be known about the species listed by clicking on their Latin bionomials. If you lack internet access we have provided a list of species from a sample country to work with (Table 3.1).

Evaluating Why Species Are Important

Now address the general question of whether it matters if these species go extinct. In other words, why are they important? In many cases you will need to do some

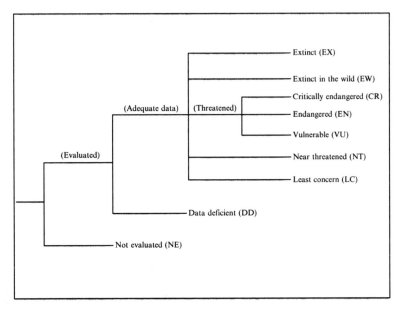

Fig. 3.1 Structure of IUCN Red List categories.

Table 3.1 A sampling of imperiled vertebrate species from the island nation of Barbados.

Latin Name	Common name (E=English, F=French, S= Spanish)	IUCN Red List category
Chelonia mydas	Green turtle (E), Tortue comestible (F), Tortue franche (F), Tortue verte (F), Tortuga blanca (S), Tortuga verde (S)	EN A2bd
Dermochelys coriacea	Leatherback (E), Leathery turtle (E), Luth (E), Trunkback turtle (E), Tortue luth (F), Baula (S), Canal (S), Cardon (S), Tinglada (S), Tinglar (S), Tortuga laud (S)	CR A1abd
Epinephelus itajara	Goliath grouper (E), Jewfish (E), Mérou géant (F), Mérou (F), Tétard (F), Cherna (S), Cherne (S), Guasa (S), Guato (S), Guaza (S), Mero batata (S), Mero guasa (S), Mero güasa (S), Mero pintado (S),	EN A2ad
Epinephelus striatus	Nassau grouper (E), Cherna Criolla (S), Cherna (S)	EN A2ad
Eretmochelys imbricata	Hawksbill turtle (E), Caret (F), Tortue caret (F), Tortue imbriquée (F), Tortue à bec faucon (F), Tortue à écailles (F), Tortuga carey (S)	CR A1bd
Guaiacum officinale	Commoner lignum vitae (E), Guaiac tree (E), Guayacán (S), Palo de vida (S), Palo santo (S)	EN C2a
Liophis perfuscus	Barbados racer (E)	EN C2b
Numenius borealis	Eskimo curlew (E), Courlis esquimau (F), Chorlito esquimal (S), Chorlo polar (S), Zarapito boreal (S), Zarapito esquimal (S), Zarapito Polar (S)	CR D

extra research on particular species to articulate their values. Reasons why a species is important can generally be grouped into the three basic categories:

- **Ecological values**: All species are supported by the interactions among other species and ecosystems. Loss of species makes ecosystems less resilient and often less productive.
- **Economic values**: Wild species provide to humans various products and services that form much of the basis for the human economy.
- **Cultural values**: The identity of human cultures around the world is attached to varying degree to wild species. Wild species are often referred to in religious texts. Outside of formal religion, many people feel connected to wild species for reasons that can be hard to explain. Some may be inspired by a species' beauty, revere it for its strength, or admire it for its cleverness.
- **Inherent values**: Values of biodiversity may also exist wholly outside the human context. Such is the case when the statement is made that all species have the right to exist and, conversely, no species deserves to be driven to extinction.

Search through the list of imperiled species of your country and identify a species that provides a significant **ecological value**, function, or service. Articulate what that ecological value is. Ecological values can often be the most elusive to identify so think hard about each species and how it fits into and contributes to the ecosystem within which it lives. Does the species provide pollination services? Does it prey on pests? Does it play a role in nutrient and carbon cycling? Is it an important herbivore? Write down the species' name and your conclusions about its ecological contribution.

Now search through the species list and identify an imperiled species that provides significant **economic value**. Articulate what that economic value is. Does the species represent a source of food? Of fiber? Of materials? Of medicine? Is it traded in some fashion? What economic loss would its extinction represent? Write down the species' name and your conclusions about its economic value.

Last search through the species list and identify an imperiled species that provides significant **cultural value**. Does the species play a role in myth? Is it beautiful? Is it scary? Is it intriguing and curious? What loss would its extinction represent? Write down the species' name and your thoughts on its cultural significance.

Evaluating Why Species are Threatened

Now answer the following general questions: What are the major threats to wild species in your country of interest? What are the major threats to species around the world? Does the intensity of threats vary substantially at the country-wide versus the global scale? If so, why?

To answer these questions you will repeatedly query the IUCN database using different search criteria. Each search will enable you to determine the number of species in a particular region that are facing a particular threat. You will then contrast the importance of each threat in your country versus globally.

Step 1 : Searching the IUCN database

Go to: www.redlist.org/ and choose Search (you can do this with an Expert Search but start with a basic search). Now under "Select one or more countries" you can limit search results to a specific country or select multiple countries. Hold down the Ctrl+key to make multiple selections. Note that the search results using the country search option exclude any species that have been introduced to the country or countries concerned (in other words, only native species are included). Choose your country or a set of nearby countries that best represents your region of interest.

Under Major Threat Types you limit your search to particular threats to species, or select one or more threats. The major threat types are numbered according to a four-level hierarchical system. Note that the threats may have been in the past, they may be having their impact at present, they may only be future threats, or they may be ongoing. Because we are interested in general classes of threats, you should search the major categories one at a time. So start with: "1. Habitat Loss/Degradation (human induced)." This will select all those species at risk from any type of human-induced habitat loss or degradation. Then select Search.

Step 2 : Interpreting the results of the search

You will now see a list of species that meet the search criteria you have selected, that is, those that occur in your country and are threatened by habitat loss and degradation. Look at the list of species. Do you recognize any of them? Do you have any personal knowledge of the specific habitat loss issues confronting any of them? Because we want to contrast threats by region, you are primarily interested in tallying the total number of species that were located by the search (see top of results screen, e.g. "Viewing results 1 to 50 of 162"). Record the total (e.g. 162 in this case).

Step 3 : Performing the remaining searches for your country

You want to obtain and record the total number of species in your country facing different threats. So repeat the search process for your country, but change only the threat category. Repeat this for the following 10 categories:

1 Habitat loss/degradation (human-induced)
2 Invasive alien species (directly affecting the species)
3 Harvesting (hunting/gathering)
4 Accidental mortality
5 Persecution
6 Pollution (affecting habitat and/or species)
7 Natural disasters
8 Changes in native species dynamics
9 Intrinsic factors
10 Human disturbance.

Record the total number of species for each threat category using Table 3.2 as a guide to organize the data.

Step 4 : Getting the global perspective

Repeat Steps 1 through 3 but now for "Select one or more countries" choose All. This will extract the total species recognized anywhere on the globe by the IUCN in relation to the type of threat. Record these totals, as you did for your country, but this time for the entire globe. Convert the numbers in each threat class to a fraction of the total species across all threat classes in your country. Do the same for the global species tallies.

Table 3.2 Sheet for tallying relative importance of threats to species at the local versus global scale.

IUCN threat category	Scale			
	Local		Global	
	Number of species	%Total species	Number of species	%Total species
Habitat loss/degradation				
Invasive alien species				
Harvesting				
Accidental mortality				
Persecution				
Pollution				
Natural disasters				
Changes in species dynamics				
Intrinsic factors				
Human disturbance				
Total species				

Expected Products

- Come prepared for discussion or a written or oral presentation with a physical description of each of the three species along with a synopsis of its natural history based on your research as well as description of their respective "values". Does it matter that they will go extinct?
- Prepare a figure portraying the relative number of species regionally (i.e. in your country or countries) versus globally in relation to each threat category and interpret the pattern in the context of the discussion questions that follow.

Discussion

1 Do we as a society bear an obligation to act as responsible stewards of the species you have examined? Should we conserve them for the present and future values that they contribute to the human species? Take sides and explore all perspectives.

One extreme position is that the fates of none of these imperiled species matter in the context of the human suffering that is so widespread today. Another extreme is that we must save every species at any cost because they all have an inherent right to exist. The bottom line is to explore whether we have an obligation to future generations to find a way to conserve these species. What do you as an individual and what does your class as a group decide?

2 Which threats are most significant to species in your country? Which threats are most important at the global level. Were there significant discrepancies between the country and the global level? If so, which threat categories were most different? What accounts for these differences? This last question is fairly complex and requires you to think of the different social, economic, legal, and biological dimensions of your country versus that of the world in general. Finally, for your country, what does your assessment of threats suggest are the most important areas on which to focus conservation actions?

Making It Happen

As you may have gathered, the IUCN database has an extraordinary wealth of information about the Earth's species diversity, its distribution, and its threats. The analysis you have just performed is cursory. Explore other questions using the database. You could, for example, restrict your analyses to particular biomes (such as marine, freshwater, or terrestrial). How about contrasting plants and animals? What is the relative risk posed by specific threats, e.g. global climate change? The possibilities are endless. Take advantage of this extraordinary database to learn as much as you can about threats to biological diversity by developing your own lines of inquiry as an individual or as a class.

Further Resources

All the major biodiversity textbooks have discussions of the values and threats to biodiversity. An excellent case study for discussion and further exploration of values is Costanza et al. (1997) and one for threats is Wilcove et al. (1998).

PART 2

Genes

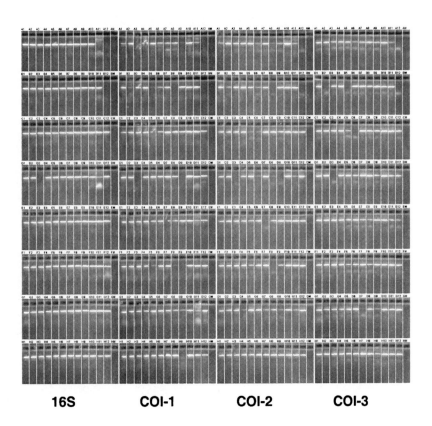

16S   COI-1   COI-2   COI-3

CHAPTER 4

Population Genetics: Diversity Within Versus Diversity Among Populations

James P. Gibbs

To contradict Gertrude Stein, "a rose is not a rose is not a rose." Any astute observer of roses knows that there is enormous variation among roses, and any person with the most basic understanding of genetics realizes that there is often enormous genetic variation among individual organisms, at least all those that reproduce sexually. Conservation biologists cannot hope to maintain all the genetic diversity represented by different individual plants and animals, but they do try to maintain the genetic variation that is represented by local populations. For example, if the roses living on one island have evolved adaptations that are different from roses of the same species living on a nearby island, it is desirable to keep both populations to maintain the full evolutionary potential of the species. Therefore, when considering how many populations to protect to keep all the genetic variation that characterizes a species, it is important to know how genetic variation is partitioned within the species. At one extreme, the genetic variation of a species can be distributed throughout all its constituent populations. In this case, populations are identical genetically and the species shows no genetic subdivision. For this species, preserving just one population will capture all the genetic variation that characterizes the species (although, obviously protecting just one population is a risky strategy given the possibility of some catastrophe). At the other extreme, the genetic variation of a species may be partitioned entirely among local populations, such that all populations must be conserved to capture the species full complement of genetic diversity. Of course, real species occur somewhere between these poles, and we need to measure the degree of subdivision or genetic differentiation that exists before we can decide how many populations need to be protected to conserve genetic diversity within a species. To do so we must understand something about the "structure" of genetic variation among populations. This requires use of analysis techniques from the field of population genetics. Thus, the purpose of this exercise is to learn to apply population genetics concepts to conservation planning.

Objectives

- To learn to "score" frequencies of genetic markers
- To learn to estimate levels of population subdivision using *F*-statistics
- To learn to draw inferences and make management recommendations on the basis of population genetic information.

Procedure

The Scenario

For this exercise, imagine that you are working for a conservation agency and are faced with a crisis decision: a piece of land with six different wetlands hosting the only known populations of two rare orchids is about to be developed. The crisis is that you only have enough funds to purchase and protect four of these wetlands. Which of the six populations should you protect? The wetlands are of two types. Three are marshes that host the world's only known populations of *Orchis isozymus*. The other three are swamps that host the only populations of the closely related but distinct *Orchis polyzymus*. Because the populations in each wetland are large (more than 1 000 plants in each) demographic issues are secondary to preserving the genetic diversity that characterizes these species. How should you allocate your scarce funds for wetland acquisition?

The Genetic Data

Unfortunately, there is no time to raise more funds. In desperation, you send some leaf samples from these plants to a colleague in a local botany department, and a week later you receive the data below, which are in the form of protein electrophoresis gels for one allozyme locus (ORCHIS1) that is polymorphic in both these species. The locus has two alleles, Fast and Slow, that appear on the electrophoresis gel with the faster allele below the slower allele (Figure 4.1).

To complete this exercise you first need to know how to interpret enzyme electrophoresis gels and the allozyme patterns that appear on them. Allozymes are enzyme variants with different electrical charges. These variants arise from small changes in the DNA sequence of protein coding genes, which in turn alter what amino acids are incorporated into proteins. When you make a slurry of tissue from an individual plant or animal, insert them into a starch gel, and subject the gel to an electrical current, the enzyme molecules with different electrical charges will migrate through the gel at different rates. This is the procedure known as electrophoresis. If after a period of electrophoresis you bathe the gel in a co-factor for a particular enzyme, as well as a dye that precipitates when the co-factor and enzyme react, then you can observe the relative mobility of each enzyme based on its position in the gel. The gels below exhibit typical banding patterns. The degree to which individuals share the same banding patterns of enzyme electrophoresis gels reflects the similarity of their genetic make-up.

Orchis isozymus, Population 1 (individual 1..15 from left to right)

	1	2	3	4	5	6	7	8	9	10	11	12	13	14	15	#	%	2pq
Slow	-	-	-	-	-	-	-	-	-		-	-	-	-	-	23		
Fast	-						-			-			-	-		7		

Orchis isozymus, Population 2 (individual 1..15 from left to right)

| Slow | - | | - | - | - | | - | | - | | - | | - | | - |
| Fast | - | - | | - | - | | - | | - | | - | | - | | - |

Orchis isozymus, Population 3 (individual 1..15 from left to right)

| Slow | - | | | | - | | | | | - | | | | | |
| Fast | - | - | | - | | - | | - | | - | | - | | - | - |

Orchis polyzymus, Population 1 (individual 1..15 from left to right)

| Slow | - | | - | | - | | - | | - | | - | | - | | - |
| Fast | - | | - | | - | | - | | - | | - | | - | | - |

Orchis polyzymus, Population 2 (individual 1..15 from left to right)

| Slow | - | | - | | - | | - | | - | | - | | - | | - |
| Fast | - | - | | - | | - | | - | | - | | - | | - |

Orchis polyzymus, Population 3 (individual 1..15 from left to right)

| Slow | - | | - | | - | | - | | - | | - | | - |
| Fast | - | | - | | - | | - | | - | | - | | - |

Fig. 4.1 The protein electrophoresis gels for one allozyme locus (ORCHIS1, polymorphic in both *Orchis polyzymus* and *O. isozymus*) returned to you by your botany colleague.

The Analysis

To measure how genetic variation is partitioned within and among populations you first need to determine allele frequencies in each population. The particular allozyme locus assayed possesses 2 alternate forms. The identity of the 2 alleles in each individual is reflected directly by the banding patterns within each lane on a gel. For example, the first individual in the first lane of the first gel (Figure 4.1) is a heterozygote, that is, the 2 alleles it possesses are different and are indicated by both a fast- and slow-moving band on the gel. In contrast, the individual in the second lane is a homozygote, as indicated by the possession of a single band representing two slow alleles. To determine the allele frequencies in each population of the fast-moving allele (p) and the slow-moving allele (q) simply tally the number of alleles across individuals and divide it by the total number of alleles present in the population (always equal to twice the number of individuals).

Next you need a measure of genetic differentiation among populations. A useful and commonly used measure is Wright's fixation index, or F_{st}, which ranges from 0, indicating no differentiation among populations, upwards, indicating increasing subdivision. Once you have calculated the allele frequencies in each population, it is relatively straightforward to determine F_{st}.

First, you need to calculate the expected heterozygosity for each species (H_s). You do this simply by multiplying $2pq$ for each population and then averaging these values over all three populations within each species. You may recall that this is the expectation of the frequency of heterozygotes in the population at Hardy–Weinberg equilibrium.

Second, you need to calculate expected heterozygosity if all three populations were part of the same, extended breeding population (H_t). You do this by averaging p and q over all populations within a species, and then multiplying 2 x the average p x the average q. This would be the expected frequency of heterozygotes in the population if it acted as one large breeding pool with no subdivision. Deviations of the frequency of heterozygotes in separate populations (H_s) from what you would expect to find if they were all part of the same larger population (H_t) provide an index of subdivision and the amount of genetic variation that is found only in local populations. Thus,

$$F_{st} = (H_t - H_s)/H_t$$

where values of F_{st} less than 0.01 indicate little divergence among populations, and values above 0.1 indicate great divergence among populations.

To complete this exercise, for each species you need to calculate the allele frequencies in each population, calculate the fixation index for each species, and then compare the indices between species.

Expected Products

- Allele frequencies (p and q) for each species and population
- Calculations of estimates of H_t, H_s, and F_{st} for each species
- Interpretation of the F_{st} values in terms of a rationale for prioritization of a particular combination of species and populations for protection.
- Responses in a form indicated by your instructor to the Discussion questions below.

Discussion

1. How will you allocate your scarce funds for wetland acquisition? Specifically, which 4 populations will you conserve? Justify your decision in terms of preserving the maximum amount of genetic diversity that characterizes these two species.
2. What are at least two causes for the different genetic structures you have observed in the two species? _potential_
3. Why choose to protect any particular population when all populations of either species have plenty of all the alleles encountered?
4. Recall that our goal in this exercise is to capture as much of the genetic diversity that characterizes these species as is possible given a limited budget. Note that both alleles at the locus surveyed already are found in each population of both species. Why does it matter if more than a single population of each species is protected?

Making It Happen

You do not need a molecular genetics lab to observe or even to quantify genetic diversity. Many species possess a great deal of local morphological variation that can be readily appreciated with the bare eye. Careful observation of many organisms often indicates subtle genetic variation among local populations (e.g. leaf morpho-metry in plants). Once you have become familiar with issues of conservation of genetic variation at the population level, can you identify any conservation issues in your local area that might benefit from a consideration of this perspective? For example, are any re-introductions of extirpated species being undertaken? Similarly, are game or fish populations being stocked or re-stocked in your area? Where are the individuals being used for these purposes originating? Has anyone considered the implications of the origins of the stocks used, or the consequences of mixing imported stocks with local stocks? You may be surprised to find that they have not. Similarly, are zoos or botanical gardens in your area collecting individuals, tissue samples, or germ plasm for conservation purposes? How representative might these samples be of the genetic variation present in the species? Have managers of the collections considered the implications of where their samples originated and the amount of genetic subdivision that may be present in the wild population? Again, you may be surprised by the answer.

Further Resources

See Frankham et al. (2002) for a useful overview of conservation genetics, and Avise and Hamrick (1996) for an interesting set of case studies.

CHAPTER 5

Genetic Drift: Establishing Population Management Targets to Limit Loss of Genetic Diversity

James P. Gibbs

Small, isolated groups tend to exhibit less genetic variability. Why is this? The main reason is a process called **genetic drift**. The smaller a population is, the more sporadically its allele frequencies "drift" around, and the higher the likelihood that one allele will be fixed at a locus while another is lost from the population. In the absence of migration or mutation, such an allele will be lost permanently. If that lost allele could have conferred some fitness advantage on individuals in the population, either at present or in the future, its loss will constrain the population's ability to adapt to its environment and persist.

The biology of small populations and the risk of "getting too small" is a key topic of every conservation biology text. We will not rehash the general issue of the problems of small populations here. Rather we will focus on an important application of concepts of genetic drift to management of wild and captive populations not well elaborated upon in conservation biology texts. More specifically, if you are concerned with maintaining a population at an effective size that permits it to retain a certain level of genetic diversity, how large should it be? To put it another way, how many individual plants or animals should be counted in the field to assure managers that genetic variation is not being drained out of it by the process of genetic drift?

The answers to these questions are surprisingly elusive. Key to answering them is the concept of "effective population size" and its relationship to "adult population size." Let's quickly define these two terms. The **effective population size** of a population is the size of an ideal population that loses genetic variability the same as the real population in question. The real population is the **adult population size**, that is, the number of mature individual plants or animals counted in the field. Thus, if a field population of 100 loses genetic variation at the same rate as an idealized population of 10, then we say it has an effective size of 10 even though it contains 100 individuals. Here's the crux of the problem: we have an elegant body of theory and supporting mathematics to estimate the effective population size needed to limit loss of genetic diversity to a particular level. But only recently have we begun to understand the relationship between the notion of an idealized notion of effective population size and the more tangible (and measurable) notion of adult population size.

The goal of this exercise is to provide you with the skills and background to make linkages between the concepts of effective and adult population size. After completing the exercise you should be able to determine, for a specified level of loss of genetic variation to be tolerated in a population, the minimum size that a population needs to be in the field.

Objective

- To learn how to link the concepts of effective to adult population sizes in order to be able to estimate numbers of individuals in the field needed to limit loss of genetic diversity to a specified level.

Procedures

Before starting the exercise it is important to review concepts of effective population sizes, rates of loss of genetic diversity, generation times, and ratios of effective to adult population sizes.

Loss of Genetic Variation as a Function of Effective Population Size

Effective population size (N_e) is a concept from the field of population genetics first defined by Wright (1931) as "the number of breeding individuals in an idealized population that would show the same amount of dispersion of allele frequencies under random genetic drift or the same amount of inbreeding as the population under consideration." The concept of an idealized population is at the heart of the well-recognized equation cited in almost every conservation biology and wildlife genetics text for estimating the loss of genetic variation over time:

$$H_t/H_o = [1 - 1/(2N_e)]^t$$

where H_t is heterozygosity after t generations, H_o is heterozygosity in the population initially, t is the number of generations elapsed, and N_e is the effective population size. Note that H_t/H_o is equivalent to the fraction of genetic diversity retained over t generations. In other words, if $H_t = H_o$ then $H_t/H_o = 1.0$, such that 100% of the initial variation remains after t generations. As an example, how much initial heterozygosity would an effective population of 25 expect to retained after 50 generations? The answer would 36%, or $[1 - 1/(2 \times 25)]^{50} = 0.364$.

The formula can be conveniently rearranged to estimate the effective population size associated with retention of a given amount of genetic variation over time, as follows (try to rearrange the formula yourself to brush up on your algebraic skills):

$$N_e = -t/[2 \ln (H_t/H_o)]$$

With this version of the formula you can ask such questions as: what size of an effective population can expect to retain, for example, 41% of its heterozygosity over 12 generations? The answer would be $N_e = 7$, or $-12/[2 \ln (0.41)]$.

Generation Time

An important point of clarification is that most biologists think of time simply in terms of years. But generation time is the key parameter here. What is generation time? Generation time is the average age at which a female gives birth to her offspring. We must have an estimate of generation time for a particular organism to convert a particular time frame specified in years to one specified in generations. For example, if we are interested in the fate of a population over 100 years and the average age at which a female gives birth to her offspring is 10 then in this case 100 years is equivalent to $t = 10$ generations. Generation times vary dramatically among species (Table 5.1) and one must think of time in a currency relevant to the organism of interest. To extend our example, 100 years represents about 22 generations of grizzly bears but only about 2 generations of white spruce (Table 5.1).

Effective Population Size/Adult Population Size Ratios

The last piece of this puzzle is that we usually only know the adult size of a population but it is the effective population size that determines the loss of genetic diversity. Thus we must know the ratio of effective population size to adult population size (N_e/N_c). Some examples of these ratios are provided in Table 5.1. The very factors that cause adult population size to differ from effective population size are those associated with making an actual population deviate from the idealized population, for example, variation in sex ratios. Most conservation biology textbooks provide the formulas for accounting for these deviations on a factor-by-factor basis. In practice, integrating all the factors simultaneously is a complex undertaking. One important review by Frankham (1995) identified the key factors as variation in population size over time primarily and variation in family size and sex ratio secondarily, with effect of generation overlap, life-history attributes, or major taxonomic groups having surprisingly little effect. An important output from Frankham's (1995) review was a summary of long-term N_e/N_c ratios among studies with the conclusion that the average was only 11% (Figure 5.1).

Table 5.1 A sampling of generation times (source for plants: primarily Franco & Silvertown 2004; for animals: primarily Martin & Palumbi 2003) and ratios of effective to adult population sizes (source: Frankham 1995) for selected species of plants and animals.

Species	Generation time (years)	Ne/Nc
Drosophila pseudoobscura (fruit fly)	0.1	0.012
Oncorhynchus tshawytscha (Chinnok salmon)	5.0	0.013
Notophthalmus viridescens (red-spotted newt)	2.5	0.073
Rana sylvatica (wood frog)	2.5	0.440
Strix occidentalis (spotted owl)	7.0	0.390
Ursus arctos (grizzly bear)	4.5	0.280
Rhinocerus unicornis (greater one-horned rhinoceros)	12.5	0.610
Peremeles gunnii (eastern barred bandicoot)	1.5	0.135
Macaca fuscata (Japanese monkey)	9.8	0.650
Picea glauca (white spruce)	50.0	0.391
Phlox drummondii (annual phlox)	1.2	0.216

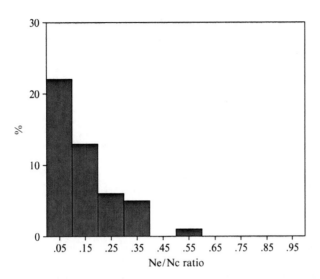

Fig. 5.1 Summary of 47 reported ratios of N_e/N_c compiled by Frankham (1995) in his synopsis. Bars represent the percentage of reports of a given N_e/N_c ratio. These are "comprehensive" estimates that factor in temporal variation in size as well as population structure. The mean ratio was 0.11, indicating that long-term effective population sizes were just 11% of adult sizes.

Pulling It All Together

The last stage in the exercise is integrating all these ideas to address some relevant conservation questions. Here's an example of the kind of problem that you can resolve: What size of adult population of grizzly bears should be maintained to limit loss of genetic variation over 500 years to less than 5%? The first step in answering this is to identify the number of grizzly bear generations involved. Looking at Table 5.1, with 4.5 years per generation of grizzly bears, then 500/4.5 = 111.1 generations of grizzly bears. Next we wish to limit loss of variation to less than 5%. In other words, we wish $H_t/H_o = 0.95$. Because we wish to solve for effective population size, we use the equation

$$N_e = -t/[2\ln(H_t/H_o)], \text{ such that } N_e = -111.1/[2\ln(0.95)] = 1\,083 \text{ bears.}$$

Note that this is the effective population size, so it needs to be calibrated by the estimated N_e/N_c ratio (Table 5.1) of 0.28. Thus, 1 083/0.28 = 3 868 bears in the field acting as a potentially interacting breeding group.

Expected Products

- The expected products for this exercise are answers to the questions posed next. To answer them, become familiar with the calculations outlined above. Note that for some species in the following questions generation times are given but N_e/N_c is not. This will often be the case in the wild, especially for rare species that we know little about. In such cases, first try to find a closely related species and use the appropriate values. If there is nothing provided that is closely related (that is, within the same genus, for example) then apply the mean ratio of 0.11 as derived by Frankham (1995) from his literature survey. If this seems incautious, consider that this is often the best we can do lacking species-specific data and Frankham (1995) found that life-history attributes or major taxonomic group had little influence on variation in N_e/N_c, that is, the ratio seems applicable for many organisms.

Discussion

1 Greater one-horned rhinoceros numbers in Royal Chitwan National Park in Nepal have been fluctuating due to restoration efforts that increase the population but also to social upheaval that has led to increased poaching levels. About 500 animals remain in the field. Is this population sufficient to maintain levels of genetic diversity at 95% of their current levels over 100 years?

2 Martinez Spruce (*Picea martinezii*) is native to northeast Mexico. It occurs at only two localities in the Sierra Madre Oriental mountains in Nuevo León where it grows along streamsides in mountain valleys. It is critically endangered. Of just two extant populations, one comprises about 300 trees and the second about 10. What is the prognosis for the amount of genetic variation that might be lost from each of these two small populations over the next 100 years? To what size should these populations be restored if 95% genetic variation is to be retained over the next 500 years?

3 *Drosophila hemipeza* is a federally listed "endangered" species that occurs only on the island of Oahu in the Hawaiian Islands where it is historically known from seven localities. Larvae feed within decomposing portions of several different mesic forest plants. The larvae inhabit the decomposing bark of *Urera kaalae* (family Urticaceae), a federally endangered plant that grows on slopes and in gulches of diverse mesic forest. In 2004, only 40 individuals of *U. kaalae* were known to remain in a single population in the wild. Assuming that each *U. kaalae* can support an adult population of some 100 *Drosophila hemipeza*, if the population of *U. kaalae* is stabilized at 40 individuals does *Drosophila hemipeza* face a significant risk of loss of genetic diversity (>10%) over the next 10 years? Does the extremely short generation time and rapid population turnover potentially buffer this fly from significant loss of genetic diversity? If not, how many *Urera kaalae* should be added to the population to build the *Drosophila hemipeza* population to a size where it no longer faces a significant risk of loss of genetic diversity (>10%)?

4 A genetic survey of an isolated population of wood frogs was conducted at two time periods. Variation in a single allozyme was assessed in the population using gel electorphoresis and the two gels, from the first survey (Year 0) and the second (Year 20) are depicted in Figure 5.2. What is the effective size of this wood frog population? How many adults do you estimate from the genetic data to be in the population?

5 The eastern barred bandicoot is a highly threatened marsupial of Australia that now survives in seven reintroduced populations. What is the minimum population size of adults in the field that should be established and maintained to ensure that drift processes in reintroduced populations will not lower levels of genetic diversity by more than 10% over the next 50 years? Habitat limitation indicates that the maximum subpopulation size is about 500; will 500 adults be sufficient to meet the goal?

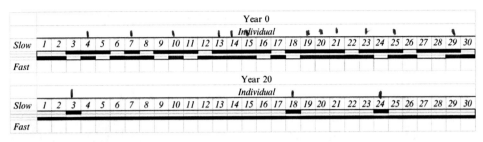

Fig. 5.2 Assays of variation in a single allozyme locus at year 0 and 20 years later in the same wood frog population. Two alleles are evident: fast and slow.

Making It Happen

Genetic drift may seem like an esoteric topic, but it is nevertheless a legitimate consideration for many common conservation programs. For example, examine the local endangered species reintroduction programs or even stocking programs for fish or game animals ongoing in your area. How many individuals are stocked and used to found the new populations? Surprisingly few individuals are often involved. Similarly, many reserves are set aside without a consideration for the sizes of the wildlife populations protected therein. What are the distributions of sizes of parks in your area? How do these correspond to the densities and, by extrapolation, population sizes of local species? Will these areas support populations of all species, especially the large-bodied ones, which are of sufficiently large effective size to buffer against drift over the long term? Raising these issues could be of real service to these efforts.

Further Resources

An excellent resource is a recent text on conservation genetics by Frankham et al. (2002). Furthermore, the "links and downloads" section the website supporting this textbook has much useful guidance: http://consgen.mq.edu.au/default.htm.

CHAPTER 6

Pedigree Management: Controlling the Effects of Inbreeding as Indicated by Fluctuating Asymmetry

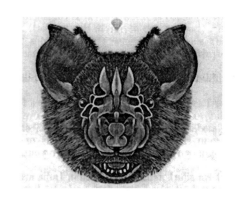

James P. Gibbs

Conservation genetics has many applications in conservation biology. It is central to successful captive breeding and subsequent re-introduction of threatened species. Many species have been saved from extinction by captive breeding in zoos, botanical gardens, and aquaria, which provide a safety net when other protective measures have failed. Yet because of costs and space limitations, bringing a species into captivity often means having to manage small numbers of individuals. Under these conditions inbreeding and genetic drift can become critical considerations. Captive management thus involves careful analysis of breeding programs and, in particular, pedigrees and monitoring of the effects of inbreeding. One external manifestation of inbreeding can be a change in body symmetry, which can be evaluated from measurement of morphological features of the individuals in captivity. This exercise focuses on these issues and seeks to integrate concepts in conservation genetics using a captive breeding program as a vehicle to do so.

Objectives

- To become familiar with the concept of inbreeding
- To learn how to construct and analyze pedigrees
- To implement morphological analysis for analyzing the effects of inbreeding
- To gain experience in making expert recommendations about captive breeding plans.

Procedures

Quick Review

Recall that the **inbreeding coefficient** is the probability that an individual has a pair of alleles that are identical by descent from a common ancestor, that is, they are

autozygous. By extension the inbreeding coefficient is also the proportion of loci that carry alleles that are identical by descent from a common ancestor. We are interested in estimating levels of inbreeding for particular individuals for several reasons. One reason is that it can tell us whether certain individuals are at risk of disease and disorders associated with lack of genetic variation. Another is that levels of inbreeding indicate whether an individual has a unique or a shared component of genetic diversity relative to other individuals in the population. This unique component is something we might want to capture (or possibly exclude) in the course of designing future breeding programs.

Building the Pedigree

The first step in estimating levels of inbreeding is to draw a pedigree that describes the relationships among all individuals (live and dead) in a population. Pedigrees systematically summarize the breeding relationships among a group of individuals (Figure 6.1).

Now let's construct a pedigree for an actual population of organisms. Consider the following summary of the breeding history within a hypothetical captive population of 11 triple nose-leaf bats (*Triaenops persicus*, Chiroptera, Hipposideridae). The species occurs along the coastal regions of East Africa and is threatened by disturbance of the caves in which it breeds, particularly by disturbance from tourists to the coral caves the bats inhabit. As you can see from Figures 6.2 and 6.3 that depict the 11 living individuals in the colony, the species is aptly named.

Here is the colony's history: it was founded in 1990 by 5 bats: a male named Bright that died in 1993, a male named Blue that died in 1995, and 2 females still alive – Fluffy and Skinny, and another male, Tiny, also still alive. All of their offspring remain alive. Bright and Fluffy have produced a male named Lucky that mated with Skinny to produce Sneaky (a female). Fluffy also mated with Blue to produce Star (female), who has subsequently mated with Tiny to produce Shy (male). Of the captive-born offspring, Lucky and Star have mated to produce a Timid (male), and Sneaky and Shy have mated to produce 3 daughters: Sleepy, Fuzzy, and Triumphant.

Construct a pedigree that represents relationships among the individuals in this breeding population, using the appropriate symbols for gender as well as indicating which individuals are alive and which dead.

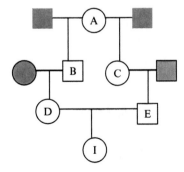

Fig. 6.1 In a pedigree, females are indicated as circles, males by squares. Offspring are indicated by lines joining their parents. This is an example of half first cousins.

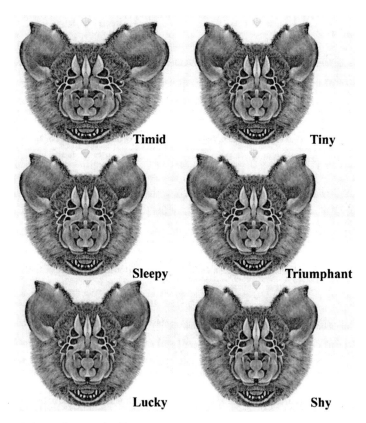

Fig. 6.2 Six of the triple nose-leaf bats (*Triaenops persicus*, Chiroptera, Hipposideridae) in the captive colony. Adapted from from Blanford 1888.

Calculating Levels of Inbreeding

The next step is to analyze the pedigree to determine the levels of inbreeding of each individual. For any given individual, we must first identify all common ancestors in the pedigree, because an allele can only be autozygous in an individual if it was inherited through both of the individual's parents from a common ancestor. Note that in many cases individuals will not have common ancestors (or at least any known common ancestors) and thus inbreeding will not be an issue for them. But for those that do, the third step is to trace all paths of gametes that lead from one of an individual's parents back to the common ancestor and then back down again to the other parent. Focusing on individual I in Figure 6.1, there is only one such path – DBACE, with A being the common ancestor (Figure 6.4).

The fourth step is to calculate the probability of autozygosity for the individuals with common ancestors. Because at each step the likelihood of passing any particular allele is ½, the probability of the same allele being reunited in a particular offspring is $1/2^i$, where i is the number of individuals in the path through the common ancestor. Last, the common ancestor may already be inbred at a level of FA. If this is the case, we add a correction factor to account for this inbred common ancestor. Thus, that the probability of autozygosity of an individual, F_I, is:

$$F_I = 1/2^i(1 + FA)$$

How inbred is each living individual in the captive colony of three-nose leaf bats for which you drew the pedigree? Note that for this captive colony, all founders

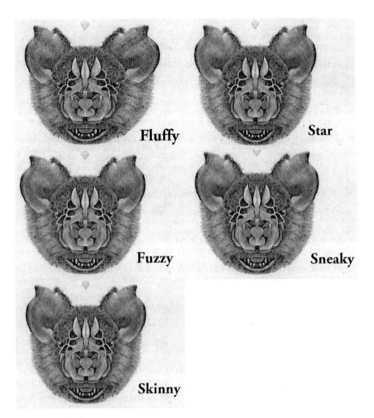

Fig. 6.3 The remaining 5 triple nose-leaf bats (*Triaenops persicus*, Chiroptera, Hipposideridae) in the captive colony. Adapted from Blanford 1888.

are/were non-inbred, except for Fluffy, who was highly inbred at $F = 0.125$. Use your pedigree to make estimates for each individual. Last, if in this species a level of $F_I > 0.1$ is known to predispose these animals to genetic disease, which individuals are at risk?

Estimating the Effects of Inbreeding by Measuring Fluctuating Asymmetry

Fluctuating asymmetry (*FA*) is deviation from perfect bilateral symmetry. It is thought that the more perfectly symmetrical an organism is, the better it has been able to handle stresses during its development because it possesses more allele variants that can buffer its development processes. Thus, individuals experiencing high levels of inbreeding often experience high levels of autozygosity and exhibit high *FA*. Low *FA* is often associated with higher mating success because *FA* is frequently assessed by potential mates during courtship (including in humans). This may occur because an individual's degree of symmetry may be an "honest" signal of one's underlying fitness. The point is that levels of fluctuating asymmetry may thus be a useful measure of overall fitness.

FA can be fairly straightforward to measure. The key concept is that we need to measure consistent parameters on the left and right side of an organism relative to the organism's midplane. In the simplest approach, the absolute value of the left-side measurement (*L*) minus the right side measurement (*R*) is an often adequate index of asymmetry. If the organism was perfectly symmetrical, then $|L–R| = 0$. If either the left or right was greater than the other, then $|L–R|$ would be more than 0. Thus,

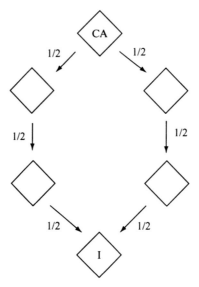

Fig. 6.4 Paths of gametes that lead from one of an individual's parents back to the common ancestor and then back down again to the other parent for pedigree in Figure 6.1.

the index starts at zero and increases positively as the individual becomes more asymmetrical.

Let's turn back to the triple nose-leaf bats. Calculate an index of *FA* for each bat. Note that you first must establish the midplane for each individual. Do this carefully with a ruler, drawing a faint line straight through the center of each individual's face. Make sure each midplane is determined as *precisely* as possible – all subsequent measurements are relative to it! Next measure on each side of the individual the same corresponding parameter. You will need a ruler with units of millimeters. Perhaps measure the distance from the center of the eye to the center of the nose for starters. But you are free to use another metric. Best is to measure several metrics and compare results among them.

Calculate your measure of *FA* ($|L-R|$) for each individual. What is your prediction about the relationship between levels of *FA* and F_i? State your prediction in writing and provide a justification for it. Now calculate the mean *FA* for inbred versus noninbred individuals. Are the estimates the same?

Expected Products

- Draw a correct pedigree for the 11 captive individuals indicating sex and status (live/dead) with appropriate symbols
- Calculate the inbreeding coefficients for each individual
- Estimate the degree of fluctuating asymmetry for each individual
- Plot the inbreeding coefficient against the estimate of fluctuating asymmetry for each individual and assess whether there is a relationship.
- Develop key recommendations for managing this breeding program in the future.
- Responses in a form indicated by your instructor to the Discussion questions below.

Discussion

Now that you have established the pedigree and learned something about levels of inbreeding in this population, let's turn the focus to broader issues. If you were to make recommendations for the breeding program in the future to capture as much as possible of what genetic variation remains in this small population, which animals would you suggest to include in future breeding efforts? Look over the pedigree carefully. Genes of which individuals are already well represented in the population? Genes of which individuals are relatively poorly represented? Consider even the dead individuals and ask in which living individuals are their genes best represented. More specifically, which individuals would you preferentially involve in future breeding programs and which would you exclude? Remember that you want to keep inbreeding levels to a minimum in whatever recommendations you make.

Making It Happen

The issues raised in this exercise are germane to any captive breeding situation, be it a zoo, botanical garden, aquarium, or endangered species breeding facility. These organizations welcome trained volunteers with new perspectives. Usually breeding programs are run under strict controls, but not always, particularly those involved in breeding of species for reintroduction to the environment, which may fall outside the purview of those familiar with pedigree management.

Further Resources

Frankham et al. (2002) is a readable and comprehensive resource on conservation genetics. For a full treatment of the topic of fluctuating asymmetry, see Palmer 1994. A major organization that links together many captive breeding interests is the Captive Breeding Specialist Group of CBSG at: http://www.cbsg.org/ Lens et al. (2002) is a good case study on application of *FA* to field conservation problems.

CHAPTER 7

Landscape Genetics: Identifying Movement Corridors

James P. Gibbs

How genetic variation is distributed within a population can reveal a great deal about the history of movement of plants and animals across landscapes. Nearly all populations exhibit genetic differentiation among geographic locales, often at a surprisingly fine spatial scale. This nonrandom distribution of variation reflects the "genetic structure" of a population that arises over time owing to the various evolutionary forces that act on a population, particularly gene flow, drift, and selection.

If the genetic structure of a population is examined using neutral markers, then natural selection can be effectively ignored as a factor structuring a population genetically. This leaves drift and gene flow, whose effects are largely antagonistic, as the main considerations in making inferences about the origins of genetic structure in a population. Where migration is limited, allele frequencies in subpopulations can drift unconstrained by gene flow, resulting in substantial differentiation among local subpopulations.

In contrast, where migration occurs, even at modest levels, gene flow mitigates differentiation promoted in local populations by drift processes and subpopulations sharing migrants become very similar genetically. Thus by mapping patterns of genetic similarity and dissimilarity among local populations and contrasting these with the make-up of the intervening environment one can begin to understand which aspects of the landscape facilitate movement of individuals and which do not. This is the essence of the new field of "landscape genetics" (Manel et al. 2003). Landscape genetics provides us with the opportunity for using population genetic studies to evaluate such important issues as landscape linkages, population connectivity, and habitat corridors. Moreover, because plants and animals can be extremely difficult and expensive to follow, the patterns in population genetic structure manifested by their exchange of genetic material are often the only way to understand how they have moved about the landscape over time.

To do these analyses you first need to characterize the genetic relationships among subpopulations. Indices of genetic similarity between population subunits are typically calculated and compiled into a matrix of genetic similarities among population subunits. Once a matrix of genetic similarity is calculated, it can be contrasted against other matrices that represent alternative models of historic gene

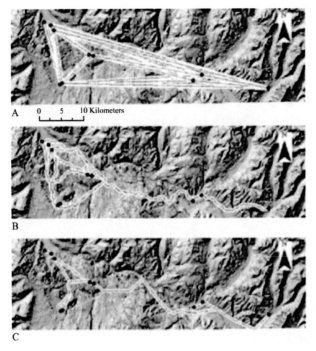

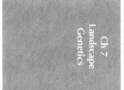

Fig. 7.1 These shaded relief maps depict three models for possible migration routes for gene flow in salamanders across a mountain range: (A) straight-line route; (B) wetland likelihood route; (C) stepping-stone route. Adapted from Spear et al. 2005.

flow within the population. For example, expectations of gene flow via riparian forest can be represented with a simple binary connectivity matrix. In such a matrix, population subunits sharing habitat connections are assigned a value of one in the matrix, and those lacking such connections a zero. Finding a relationship between genetic similarity and some measure of landscape connectivity suggests that genes (and hence individuals) have likely flowed along those routes in the recent past. An example of various models of gene flow for salamanders is provided in Figure 7.1.

In this exercise, you will apply the concepts of landscape genetics to identify how a hypothetical, rare species of a wetland-breeding amphibian – a rare toad – disperses through the larger landscape. The toad is restricted to small, isolated wetlands in an otherwise hilly and dry landscape. You wish to know how the toad moves about the environment so that you can protect its migration corridors. The two main possibilities that you will test are: (i) overland dispersal by adults or (ii) dispersal by larvae or adults via watercourses. Resolving this issue will greatly increase your understanding of the species' biology and what you can do for it in terms of comprehensive habitat protection.

Objectives

To learn how to:

- Calculate genetic distance between populations
- Develop a matrix of genetic distance
- Represent different possible models of gene flow through the landscape
- Make inferences about the relative importance of various landscape features for structuring a population genetically.

Study Area and Major Question

A map of your study area is indicated in Figure 7.2. The 6 local populations of the toad that have been genotyped are indicated on the map.

Genetic Data

Allozyme electrophoresis was used to examine the genetic structure of this population. The individual genotypes of a sample of 30 individuals from each of the 6 local populations are depicted in Figure 7.3.

Calculating Genetic Distances Among Pairs of Populations

There are many measures of genetic distance. In this exercise we will use a genetic distance estimator from Schweiger et al. (2004).

Step 1 : Calculate p and q

To estimate F_{st}, you first need to calculate p and q – the frequency of the two alleles in each population. Remember that there are 30 individuals each with 2 alleles, so p = "slow" alleles/60 and so q = "fast" alleles/60. For example, for population 1, $p = 0.90$ and $q = 0.10$. Calculate these for the other local populations.

Step 2 : Calculate H_t

Recall that $H_t = 2(p_{average})(q_{average})$ where $p_{average}$ is the average of the p values in each of the 2 populations and $q_{average}$ is the average of the q values in each of the two populations. Because there are 6 populations there are 15 unique pairs of local populations to make this calculation for.

Step 3 : Calculate H_s

Recall that H_s = average of the $2pq$ values from the 2 local populations. So determine $2pq$ for the first, $2pq$ for the second, and then average the two values. Here, too, there are 15 unique pairs of local populations to make this calculation for.

Step 4 : Calculate F_{st}

Recall that $F_{st} = (H_t - H_s)/H_t$. You already calculated H_t and H_s in steps 2 & 3 so now calculate F_{st} for the 15 unique pairs of local populations.

Step 5 : Calculate genetic distance

The last step is to transform the genetic distance values $F_{st}/(1 - F_{st})$ following Schweiger et al. (2004). This transformation simply accentuates the differences between the smaller versus larger values of F_{st}. Go ahead and do this for the 15 unique pairs of local populations.

Assemble the $F_{st}/(1-F_{st})$ values into matrix of genetic distance among local populations. Do this by filling in the elements on the lower left side of Table 7.1.

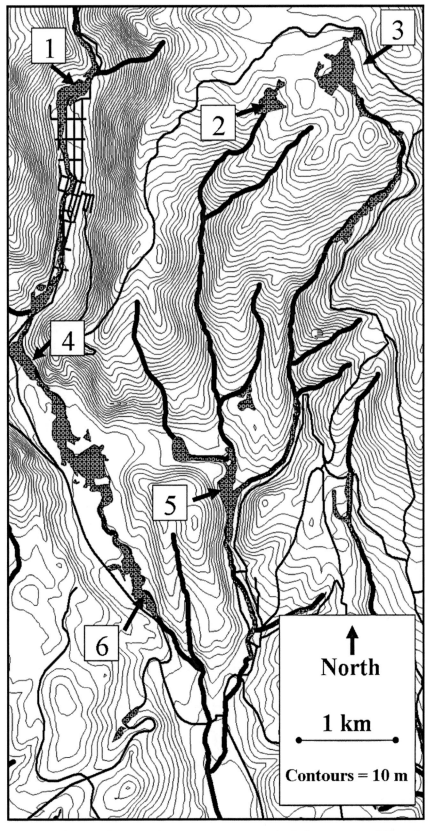

Fig. 7.2 Map of the study area. Elevation contours are the light gray nested lines. Streams are indicated with the heavy, dark lines. Roads are indicated with the light, dark lines. Wetlands are indicated as darker polygons.

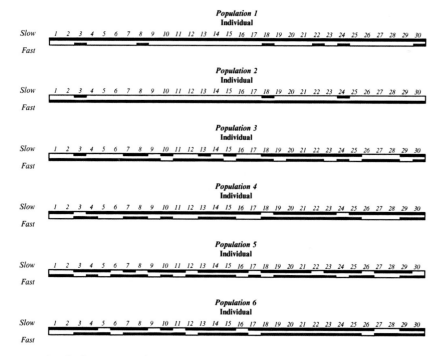

Fig. 7.3 Individual genotypes for a sample of 30 individuals in each of 6 local populations of a rare, wetland-associated toad which breeds at the locations depicted in Figure 7.2.

Make the Geographic Distance Matrix to Test for Overland Dispersal

Recall that the two main predictions you wish to make are that: (i) dispersal is primarily via the species' overland movement, or (ii) dispersal is primarily via the species moving along stream channels. Relative to other amphibians, toads are fairly dessication resistant and might possibly move overland as easily as along stream channels to get to their breeding areas. In the first case, how might you represent the process of dispersal via overland dispersal? Although it is not a perfect representation, given variation in the topographical relief of the study area, you can assume that simple geographical proximity might be a good model. In other words, if toads move in many directions overland, then local populations located

Table 7.1 Genetic distance among the six breeding populations of toads.

	Genetic distance ($F_{st}/(1 - F_{st})$)					
	Pop. 1	Pop. 2	Pop. 3	Pop. 4	Pop. 5	Pop. 6
Pop. 1	–	–	–	–	–	–
Pop. 2		–	–	–	–	–
Pop. 3			–	–	–	–
Pop. 4				–	–	
Pop. 5					–	–
Pop. 6						–

Table 7.2 The matrix of geographical distances between pairs of local populations to represent the overland gene-flow model via terrestrial adults.

	Geographic distance matrix (km separating each)					
	Pop. 1	Pop. 2	Pop. 3	Pop. 4	Pop. 5	Pop. 6
Pop. 1	–	–	–	–	–	–
Pop. 2		–	–	–	–	–
Pop. 3			–	–	–	–
Pop. 4				–	–	–
Pop. 5					–	–
Pop. 6						–

nearer to one another will be more similar genetically than those farther apart simply because the frequency of exchange of individuals will decline with distance. Thus, you need to construct a matrix of geographical distances between pairs of populations. Use a ruler or string to measure these distances. It is simplest to just measure the distances with whatever measuring device you have, then measure the 1 km scale on Figure 7.2 with that same device, then divide all your measures by the length of the 1 km scale. This will convert all measures to units of 1 km. Table 7.2 is provided to help organize the data.

Make the "Streams Distance" Matrix to Test for Dispersal Along Streams

In the case of dispersal primarily via the individuals moving along water channels, local populations linked more closely by streams should be more genetically similar than those linked more distantly by streams. A good representation of this is a "streams distance" model (Table 7.3). To develop it, you should find the shortest

Table 7.3 The matrix of distances along streams between pairs of local populations to represent the active, water-dispersed gene-flow model via adults or larvae.

	Streams distance matrix (km along connecting streams)					
	Pop. 1	Pop. 2	Pop. 3	Pop. 4	Pop. 5	Pop. 6
Pop. 1	–	–	–	–	–	–
Pop. 2		–	–	–	–	–
Pop. 3			–	–	–	–
Pop. 4				–	–	–
Pop. 5					–	–
Pop. 6						–

distance between each pair of populations *along stream courses*. Again, a string or map measurer should be useful for determining these distances. Note that the geographic and streams distance models are quite distinct because some populations that are proximal in terms of horizontal distance, e.g. 1, 2, and 3, are separated by long distances if one travels along streams only.

Correlate the Genetic Distance Matrix to the Geographic and Streams Distance Matrices

The last step is to correlate the genetic distance matrix to the geographic and streams distance matrices. This will permit you to test which model better explains the genetic distance between local populations. Note that in both cases we would expect that genetic distance will be low when distance (geographic or along streams) is low, so a positive correlation is expected in both cases if either model is relevant. To correlate the matrices, plot the values of genetic distance measure for a given pair of populations with their respective distance measure for a given model (geographic distance or streams distance). There are 15 unique pairs of local populations so there will be 15 data points to plot against one another. You may also wish to use a correlation coefficient to measure the strength of these relationships although a simple plot will suffice to indicate the general patterns. Note that the values from the environmental matrices should be your independent variable (x) and the genetic distance values the dependent variable (y). Which model describes the stronger relationship: overland dispersal or dispersal along water courses?

For advanced students   If you wish to calculate the statistical significance of the genetic distance-environment relationships note that these data violate a fundamental assumption of most statistical tests, that is, independence of data. The reason is that each local population is in fact represented in the data set five times, that is, with each possible unique pairing of it to the other populations. If one local population is quite skewed genetically, its genetic distance will tend to be extreme in all five comparisons and this will generally skew the genetic distance estimates in aggregate.

There is a way around this problem, and it involves a common technique used in landscape genetics. The solution is a so-called Mantel test, which randomizes the data to estimate significance, and thereby accommodate the problems of lack of independence within the data by comparing it to random combinations of itself. These random combinations will retain the correlated structure of the entire data set. We have designed Tables 7.1, 7.2, and 7.3 so that if you wish to perform Mantel tests you can easily do so with the "free" PopTools add-in to MS Excel, available from www.cse.csiro.au/poptools/.This is a very useful set of spreadsheet tools that is easily installed. You will find the Mantel test under the Extra Stats submenu. Directions are easily followed, especially if you first examine the demo of the Mantel test under Demos.

Expected Products

- Calculations of p and q for each local populations
- Calculation of H_s, H_t, F_{st} for each unique pair of local populations

- Calculation of genetic distance between each pair of local populations and assembly of these estimates into a genetic distance matrix
- Estimation and development of the geographic distance matrix
- Estimation and development of the streams distance matrix
- Plot of genetic distance versus geographic distance for each unique pair of local populations and an estimate of the associated correlation coefficient
- Plot of genetic distance versus stream distance for each unique pair of local populations and an estimate of the associated correlation coefficient
- Advanced students may wish to estimate the significance of each correlation coefficient
- Inference about what is the most likely means of dispersal by this toad in the landscape.
- Responses in a form indicated by your instructor to the Discussion questions below.

Discussion

1 What is the most likely mode of dispersal of this species? Why do you so conclude?
2 Given your results and conclusions, what is the most logical direction for conservation strategies to maintain population connectivity in this developing region?
3 Clearly we have not explained all the variation in genetic relatedness among these local populations... What other hypotheses might you develop for this species and the map provided? How might you represent these quantitatively as environmental matrices?
4 Why is it important to base the genetic surveys on selectively neutral markers?
5 Why in the final analysis are the environmental matrix values your independent variable and the genetic distance values your dependent variable?
6 Genetic structure is largely the product of historical patterns of movement. Do you see problems with using this approach to tackle contemporary conservation issues?
7 Are hypotheses and models of gene flow likely to be mutually exclusive? How might you deal with this issue in analyzing your data?
8 How might you apply this concept to a species of conservation concern and of interest to you? What are the compelling questions about its patterns of dispersal? Assuming you could obtain the appropriate genetic data, how would you represent the different environmental models of movement?

Making It Happen

Genetic data on population structure can provide a long-term perspective on the importance of landscape linkages. This perspective can complement the more contemporary perspective provided by behavioral studies of dispersing animals. Empirical studies have revealed many instances of strong associations between the physical structure of landscapes and the genetic structure of populations at the spatial scale of landscapes. These studies suggest that combining modern genetic techniques with matrix-congruence analytical methods could provide powerful insights into the utility of particular landscape linkages for wildlife populations. This kind of work is definitely more in the realm of conservation science than conservation management. This said, if you have interests in learning how plants and animals move about the landscape this is an interesting area to explore.

Further Resources

There are many new papers in this rapidly developing field; a good overview is Manel et al. (2003) and two excellent case studies are Hirao and Kudo (2004) for plants and Geffen et al. (2004) for animals.

PART 3

Populations

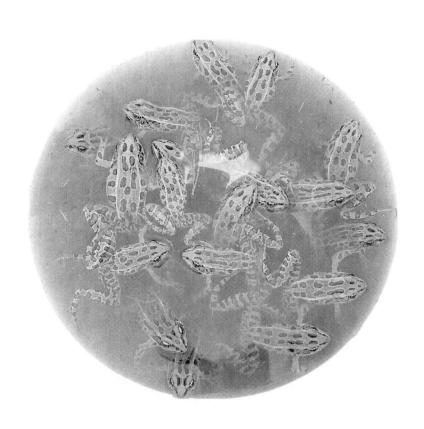

CHAPTER 8

Life Table Analysis: Balancing Commercial Fisheries with Sea Bird "By-Catch"

James P. Gibbs

Debates over conservation issues often focus on population management issues. These issues may be how to reverse declines in endangered species or how best to harvest in a sustainable fashion populations of abundant species. Any population undergoes changes through time that are determined by interactions between age-specific mortality and fecundity rates and the numbers of individuals of different ages. A first step toward understanding these processes in a population for species in which individuals can be accurately aged is to use life table analysis to calculate the population's vital statistics. Many insights into the population's behavior, and into management options for that population, can subsequently be obtained.

This exercise aims to provide you with an introduction to applied demographic analysis. We examine a hypothetical scenario involving seabirds and harvest of fishes in a marine reserve. Our goal is to balance simultaneously sustained yield in terms of direct harvest of the fishes and incidental take of the seabirds. The exercise is intended to provide you with first-hand experience analyzing the type of population data typically gathered by field biologists and to do so within an unfortunately realistic context.

Objectives

- To determine a population's vital statistics from a set of field observations
- To use these statistics to make informed decisions about ways to manage harvest of two populations.

Procedures

The Scenario

This scenario is contrived but matches closely many details of conflicts between commercial fisheries and seabirds. It is patterned after the demography of a long-lived and particularly valuable fish, such as tuna or swordfish subject to long-line fishery, and surface diving seabirds, such as gannets attracted to baited hooks associated with the long-line fishery. Consider a hypothetical species of small seabird that matures at age 3 and then breeds at a modest level (2 or 3 chicks per year) for approximately 3 years. The seabird typically forages on a fish species avidly sought after by commercial fishers. We are concerned with a 10,000 km^2 marine reserve. The seabirds typically occur at a density of 1 individual (of all ages) per 10 km^2, although seabird densities can reach twice that level in terms of the carrying capacity of the marine reserve ($K = 2$ seabirds/10 km^2). In contrast, the fishes occur at some 100 per km^2, although under the best of conditions their numbers can reach twice that, that is, at their carrying capacity (K). During fish harvests long-line hooks will incidentally kill seabirds that mistakenly pick up the hook and drown. The rate at which this happens is, on average, 1 seabird for every 1 000 fish hooked.

Based on this scenario the exercise tackles three questions:

1 How many individuals can be sustainably removed from the seabird population each year?
2 What is the maximum sustained yield of the fishes?
3 Given that some seabird mortality occurs that does not relate to the fish harvest, is managing fish harvest at its maximum yield level acceptable in terms of maintaining the seabird population?

Observations from the Field

Longevity data were collected from the field from an undisturbed population of the seabirds (Table 8.1). Listed are longevity data from a cohort of 91 individual seabirds and fecundity data from a sample of 25 females at a nesting colony. From these data you need to determine the number of individuals that survived to each age, starting with the original 91 at age 0.

The next set of data describes the maternity rates in the seabird population (Table 8.2). The data were collected from a representative sample of females in a nesting colony. Biologists checked nests of known-aged females and determined how many young they produced. Each individual is just one sample of the reproductive output of a female of a particular age. You will note from the data that reproductive maturity is reached at age 3 years and senescence sets in by age 6 years. The maternity data correspond to the number of offspring (of both sexes) produced by females of a particular age. From these data you need to calculate the average maternity rate for females of a given age. We assume that reproduction in males follows a similar pattern with age.

Construct the Life Table

Now construct a life table (see Table 8.3) for this population so that you can determine the (i) survivorship values, (ii) survival rates, (iii) mortality rates, (iv) maternity rates, and (v) population growth rates.

The first step is to determine the number of individuals alive N_x at the beginning of each time interval x. N_0 will equal the total number initially present in the cohort, or 91. For the subsequent time intervals, determine the number of individuals still alive from the longevity data provided above. The easiest way to do this is to tally

5-丗|||| 6-丗 7-|||| 8 -|

Table 8.1 Longevity data for 91 seabirds. The data are the age at death (in years) of a sample cohort member. For example, "#91 dead at 0.1 years" represents individual #91 that died at age 0.1 years (or about 1 month).

#1 dead at 4.2 years	#11 dead at 1.0 years	#21 dead at 6.4 years ||||
#34 dead at 3.3 years	#44 dead at 2.0 years	#54 dead at 2.2 years
#67 dead at 8.8 years		#77 dead at 7.4 years |||	#87 dead at 2.2 years
#2 dead at 5.0 years		#12 dead at 2.2 years	#22 dead at 0.1 years
#35 dead at 5.0 years ||	#45 dead at 0.2 years	#55 dead at 0.1 years
#68 dead at 0.7 years	#78 dead at 1.0 years	#88 dead at 4.4 years
#3 dead at 5.0 years|||	#13 dead at 1.0 years	#23 dead at 1.2 years
#36 dead at 0.7 years	#46 dead at 0.1 years	#56 dead at 0.2 years
#69 dead at 7.6 years|	#79 dead at 6.2 years|	#89 dead at 1.0 years
#4 dead at 0.9 years	#14 dead at 1.2 years	#24 dead at 3.0 years
#37 dead at 2.9 years	#47 dead at 5.0 years ||||	#57 dead at 1.0 years
#70 dead at 1.2 years	#80 dead at 5.0 years 丗	#90 dead at 3.2 years
#5 dead at 2.3 years	#15 dead at 6.0 years ||	#25 dead at 1.2 years
#38 dead at 7.3 years||	#48 dead at 0.9 years	#58 dead at 3.0 years
#71 dead at 1.0 years	#81 dead at 2.2 years	#91 dead at 0.1 years
#6 dead at 3.7 years	#16 dead at 0.1 years	#26 dead at 5.0 years 丗 |||
#39 dead at 2.7 years	#49 dead at 3.3 years	#59 dead at 0.9 years
#72 dead at 4.0 years	#82 dead at 0.1 years	#27 dead at 1.9 years
#7 dead at 3.2 years	#17 dead at 4.1 years	#60 dead at 0.3 years
#40 dead at 2.2 years	#50 dead at 0.7 years	#28 dead at 1.6 years
#73 dead at 3.9 years	#83 dead at 7.0 years 丗	#61 dead at 4.7 years
#8 dead at 1.4 years	#18 dead at 0.3 years	#29 dead at 5.7 years 丗 ||||
#41 dead at 2.4 years	#51 dead at 5.2 years 丗		#62 dead at 0.2 years
#74 dead at 3.3 years	#84 dead at 6.9 years |||	#30 dead at 6.8 years 丗
#9 dead at 1.0 years	#19 dead at 2.0 years	#63 dead at 2.4 years
#42 dead at 4.0 years	#52 dead at 0.4 years	#31 dead at 1.2 years
#75 dead at 0.7 years	#85 dead at 0.3 years	#64 dead at 0.1 years
#10 dead at 0.2 years	#20 dead at 0.1 years	#32 dead at 0.4 years
#43 dead at 0.2 years	#53 dead at 1.0 years	#65 dead at 2.2 years
#76 dead at 3.2 years	#86 dead at 5.7 years 丗 ||	#33 dead at 0.9 years
		#66 dead at 0.1 years

Table 8.2 Maternity rates in the seabird population. For example, 3 at 5 years indicates that 3 offspring were produced by this 5-year-old female.

2 offspring at 3 years	6 offspring at 4 years
0 offspring at 4 years	0 offspring at 5 years
1 offspring at 5 years	0 offspring at 3 years
1 offspring at 3 years	1 offspring at 4 years
1 offspring at 4 years	2 offspring at 3 years
2 offspring at 3 years	2 offspring at 3 years
3 offspring at 3 years	3 offspring at 3 years
4 offspring at 4 years	3 offspring at 5 years
4 offspring at 4 years	3 offspring at 4 years
3 offspring at 5 years	4 offspring at 5 years
3 offspring at 3 years	4 offspring at 4 years
3 offspring at 4 years	3 offspring at 3 years

3
10

4
9

5
5

Table 8.3 Useful format for organizing your life table calculations.

x	n_x	l_x	s_x	d_x	m_x	Other calculations
0	91					
1						
2						
3						
...						

the number of individuals that died at each age. Then make a cumulative summation from the oldest age class upward to the youngest. This will provide you with each of the N_x values, with $N_0 = 91$ as noted earlier. So, if the last individual died at age 8, then $N_8 = 1$, if 4 others died at age 7, then $N_7 = 4 + 1 = 5$, if 5 others died at age 6, then $N_6 = 5 + 4 + 1 = 10$.

Survivorship values are denoted by the symbol l_x and are the proportion of individuals born who survive to an age x. The 0th survivorship l_0 is defined to be 1 (all individuals born are alive). Thereafter the survivorship values get smaller until they reach 0 at the maximum age. Thus survivorship at any age x is simply the proportion of the 91 individuals born that are still alive at that age. So, given the example N_x values above, $l_8 = N_8/N_0 = 1/91 = 0.011$, $l_7 = N_7/N_0 = 4/91 = 0.045$, etc.

Survival rates, s_x, are the proportion of individuals alive at age x that will survive to age $x + 1$. The survival rate of age x is equal to the quotient of l_{x+1}/l_x. Survivorship values decrease with age and survival rates will range from 0 to 1. So, given the sample l_x values above, $l_7 = 0.011/0.045 = 0.24$, etc.

Mortality rates, d_x, are the proportion of individuals alive at age x that will die before age $x + 1$. These are the reverse of survival rate values, so for any age x, $d_x = 1 - s_x$.

The maternity rate, m_x, is the number of individuals produced per unit time per individual of a given age. Calculate the average maternity rate for individuals of each age from the maternity data provided above.

Calculate Vital Statistics for the Seabird Population

From the life table you have constructed (following Table 8.1), you can calculate useful statistics for the population. The first such statistic is the **net reproductive rate**, which is a measure of the productivity of a population:

R_0 Net reproductive rate = average number of female offspring that will be produced per female during her lifespan = $\Sigma l_x m_x / l_0$, or $\Sigma l_x m_x$ when $l_0 = 1$

This is calculated by summing the products of the age-specific survivorship and maternity values in the life table. Simply multiply the values of l_x and m_x across the rows of the life table and then sum these products over the various age classes. The sum will equal the net reproductive rate for the population.

What is the Mean Generation Time of This Population?

A second useful statistic is the **generation time** (G):

Mean generation time = average time between the births of a female and the birth of her offspring = $(\Sigma x l_x m_x)/R_0$

Simply multiply the values of x (the age class) and l_x and m_x across the rows of the life table and then sum these products over the various age classes and then divide by R_0.

What Is the Estimate of the Rate of Population Growth? Does the Population Appear to Be Increasing or Decreasing?

A third useful statistic is the **intrinsic rate of increase**. An approximation of this rate is:

$r_{estimate}$ the intrinsic rate of increase, which is in essence the net reproductive rate adjusted for generation length to provide a measure of population growth per unit of time, that is, $\ln(R_0)/G$.

What Is the Maximum Harvest Rate for the Seabird Population?

Now calculate what the maximum sustained yield for this seabird population might be. Recall that the equation for maximum sustained yield is $r_{est}(K/4)$, where K is the size of the population when it is unharvested, that is, at carrying capacity.

Evaluating a Harvest Strategy for the Fishes

Rather than have you develop the life table from raw field data for the fishes as you have already done for seabirds, it is provided for you in Table 8.4. Look it over. You will notice that fishes have a very different life history than do seabirds. Note also that this life table is for both sexes, so no adjustments need to be made for different sexes. The fishes take somewhat longer to mature but have vastly higher maternity rates than do the seabirds, but the fish offspring are heavily depredated by many organisms in the sea and hence have very low survival rates. However, if they do survive their first year they enjoy fairly high survival rates to grow to maturity at 15 years, when they begin to produce large numbers of young at a rate of some thousand offspring per individual per year for 10 more years before the

Table 8.4 Age distribution for the fishes. Age-specific maternity rates are given in the text.

X	m_x	X	m_x
0	10000	13	64
1	100	14	61
2	97	15	58
3	94	16	55
4	91	17	52
5	88	18	49
6	85	19	46
7	82	20	43
8	79	21	40
9	76	22	37
10	73	23	34
11	70	24	31
12	67	25	28

mature fishes succumb. Examine this life table, and calculate the vital statistics of the fish population.

Now determine the values of R_0, G, and r_{est} for this species of fish. Then determine the total annual maximum harvest of fishes from the entire reserve.

Balancing Fish Harvest Versus the Incidental Drowning of Seabirds

Now let's first evaluate whether harvesting fishes at their optimal level can be tolerated by the seabird population, which is incidentally affected when long-line hooks for the fishes inadvertently drown the seabirds. What is the estimated maximum sustainable annual harvest rate for the fishes? What is it for the seabirds? Based on data originally given in the scenario, recall that the area is $10\,000\,km^2$, that $10\,km^2$ hosts 1 seabird (of any age) and 1 000 fishes, that deploying long-line gear kills 1 seabird for every 1 000 fish it secures, and that carrying capacities for each species are twice their current densities. You now have all the information you need to address the primary question that is the crux of the exercise: If the fish harvest is set at its maximum sustainable limit, will the seabird population be secure?

Expected Products

- Completed life table for the seabird
- Estimates of the net reproductive rate, mean generation time, intrinsic rate of increase, total annual maximum sustained by-catch of the seabirds from the entire reserve
- Completed life table for the fish
- Estimates of the net reproductive rate, mean generation time intrinsic rate of increase, total annual maximum sustained by-catch of the fish from the entire reserve
- Assessment of whether the fish harvest, if set at its maximum sustainable limit, will jeopardize the seabird population
- Responses in a form indicated by your instructor to the Discussion questions below.

Discussion

1 What are the key assumptions involved in making inferences about populations using data and methods such as these? Specifically, what assumptions are you making about the field observations you have collected? Also, what assumptions are you making about the population and the stability of the environment in which it occurs?

2 How might you apply estimates of reproductive values to management of a particular population?

3 In which species, the seabird or the fish, would you expect to see the most rapid responses to conservation measures?

4 How might you use estimates of generation times in making conservation decisions about a particular species?

5 What would be the advantages of including a series of sanctuaries where no harvests can occur in a fisheries management area?

65

Life Table Analysis

Making It Happen

Life table analysis, most often performed by insurance adjusters, is quite a technical undertaking, but the insights it can provide for biologists into a population's dynamics and composition can be valuable. Constructing a life table for a given organism also forces you to think critically about the various life stages an organism passes through and the various constraints it faces at each of those stages. This can in turn provide you with a more enlightened perspective when entering into discussions of management of particular populations. Many contemporary wildlife conservation debates center in one way or another on population management issues. Consequently an understanding of life table analysis can make you a much more enlightened participant in these debates.

Harvest management and quota setting in particular are complicated undertakings. This is particularly true in fisheries management where the economic stakes are very high. Also, because revenues from hunting-related activities contribute substantially to the income of many wildlife agencies, managing game species is a major preoccupation of these agencies. A useful further exploration of this topic is to obtain a list of the species permitted to be hunted and fished in your area. You might be quite surprised at the species you find there. Similarly, examine the "bag" limits on take and the seasonal restrictions on "take" for each species, These often vary strikingly among species. Based on the biology of these species, is it intuitive why these limits have been placed? It can be revealing to discuss with agency personnel what kind of data they rely upon to manage harvests and how they set quotas for upcoming harvest seasons. Sometimes it is merely guesswork. Analyses such as these can help make the process more scientific.

Further Resources

A highly readable treatment of life table analysis can be found in Gotelli (2001) who also introduces concepts of demographic analysis for stage-structured populations, which represents a more sophisticated and extremely flexible approach to demographic analysis in a variety of organisms. Freeware packages for estimation of survival rates from field observations of marked organisms (e.g., SURVIV, JOLLY, JOLLYAGE, SURGE, MARK) can be found at a software archive of the National Biological Service: www.mbr-pwrc.usgs.gov/software.html.

CHAPTER 9

Population Viability Analysis: El Niño Frequency and Penguin Population Persistence

James P. Gibbs

"Life is tough; just when you get everything figured out, somebody goes and changes all the rules." With a bit of anthropomorphism, this very human lament could be applied to most species because survival in an ever-changing world is difficult. Indeed, we can say with considerable certainty that just as death is the fate of every individual, extinction is the ultimate fate of every population and species. But this is not a call for pessimism. Just as we strive to prolong our lives with good nutrition, medical care, and so on, we can work to stem the tide of extinctions that human activities have generated. A first step is trying to predict the effects of our actions on the fate of particular species in a changing and variable world. This is a surprisingly difficult task. Many variables interact to determine the size of a population and how long it might persist. Often the only way to gain insight is to develop a mathematical model of the population and use it to perform a Population Viability Analysis, or PVA.

Population viability analyses are based on models that relate a dependent variable (such as population size) to the independent variables that influence it (such as weather, harvest levels, mortality, etc.). The relationships between independent and dependent variables are mediated through the model's parameters (such as survival rates and reproductive rates of individuals). A PVA uses a model that combines both the magnitude of the parameters and the amount that they vary. It generally involves three steps. First, a single population projection is made over a specified time period (typically 10–100 years). The population size at any given time step is a function of both the population size at the previous time step and of values drawn at random. These random values come from statistical distributions that describe the pattern of variability in the model's parameters. Second, many such projections are made (typically 500 or more). Finally, the proportion of projections in which the population reached a certain threshold is determined. Thus, a prediction from a PVA generally has three elements: a population threshold (often 0), a probability (from 0 to 1, or 0% to 100%) that the population will reach that threshold, and an interval of time for which the prediction pertains. PVA is a form of risk analysis in which one is interested not only in the average size of a population at some time in the future, but perhaps more importantly in the range of possible future values. Therefore we make many population projections and look at what percentage of those projections fall above or below a certain population level. For example, the risk of extinction would be 37% if 370 of 1 000 population projections reached a threshold of 0. This is how extinction risk is estimated.

Objective

- To understand the elements of PVA by performing one using standard spreadsheet software with simple conditional statements to address a question concerning the fate of a small population in an uncertain environment.

Procedures

Background

Here we develop and apply a simple PVA to explore the likely fate of a fascinating and unusual species perched on the very limits of its range of physiological tolerance: the Galápagos penguin (*Spheniscus mendiculus*). Here's the issue: the penguin lives in a place rich in food resources but susceptible to occasional food shortages associated with the well-known El Niño phenomenon. In normal years, the waters of the Galápagos are bathed in cold currents with fish aplenty for the penguins. But in El Niño years, which have historically occurred every 5–7 years, warm waters enter the area and fish numbers decrease. The penguins suffer from food shortage but also the associated heavy rainfall that floods their nests. There is a very strong relationship between annual population change in penguins and change in sea water temperature in Galápagos (Figure 9.1) such that following El Niño events penguin populations drop by up to 77%. As you can see from Figure 9.1, during normal years the penguin population grew on average by 18.8% (a factor of 1.188) but during El Niño years the population crashed by an average of −69.0% (a factor of 0.310).

One prediction of global climate change is that the frequency of El Niño events will increase (indeed, there is strong suggestion that it already has, Root et al. 2007). Currently it seems that the penguin population is able to recover after periodic El Niño events but can it manage to do this repeatedly if the frequency of El Niño events shifts from its historic occurrence of every 5–7 years to every 2–3 years? The answer is not at all intuitive. This is exactly the kind of question PVA was developed to address. Here we develop a PVA in a spreadsheet environment. We use this approach to avoid the "black-box" nature of much population simulation software. We strongly encourage you to explore some of the sophisticated PVA software (see Further Resources), but we suggest you complete this exercise first so you can understand the inner workings of more complex PVA applications.

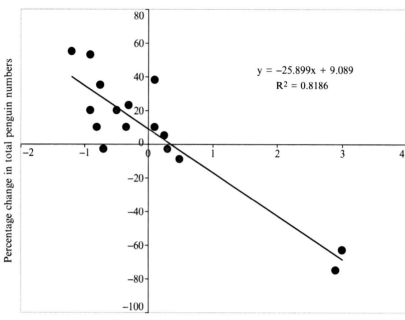

$y = -25.899x + 9.089$

$R^2 = 0.8186$

Fig. 9.1 Change in Galápagos penguin populations in relation to sea surface temperature change in the period that preceded the penguin census. Adapted from Root et al. (2007).

Modeling the Dynamics of the Penguin Population in a Simple Spreadsheet

A surprisingly realistic model of the dynamics of this penguin population can be developed as a simple spreadsheet exercise. First let's have a conceptual overview of how the model works.

Step 1 : Will it be an El Niño year?

Historically they have occurred every 5–7 years. For simplicity sake, let's say every 7 years as the baseline. Thus, the probability of an El Niño event occurring in any given year is 1/7 or 0.14. In terms of modeling these El Niño events through time, we just need to generate for a given year in the spreadsheet what's called a "uniform random number," which is a random draw of a value between 0 and 1 in which any value is equally likely. Here's a sample of 10 uniform random numbers: 0.45, 0.84, 0.27, 0.71, 0.51, 0.99, 0.05, 0.82, 0.98, 0.59. If the uniform random number was below the annual chance of an El Niño event, 0.14, then we'd use the population growth for an El Niño year (the change factor of 0.310). If the uniform random number was above the annual chance of an El Niño event, 0.14, then we'd use the population growth for normal year (the change factor of 1.188). In this case, only one of 10 years was designated as a El Niño event.

Step 2 : What will be the size of next year's population?

The next step is to project the population from one year to the next for some extended period. Let's use 100 years. You will start with 500 penguins. The number the following year will be a function of the number you started with (500) times the population growth rate. Again if it's an El Niño year you would multiply 500 by 0.310 but if it's a normal year you would multiply 500 by 1.188. You have now projected the population from year 0 to year 1. From here on you just repeat the process, multiplying the population in year 1 by the appropriate growth rate depending on if it's an El Niño year or not to get the population in year 2, then doing the same to project the population from year 2 to year 3, and so on through to year 99 (we'll stop at year 99 because we started at year 0).

Step 3 : Has the population exceeded carrying capacity this year?

The last thing we need to do is to ask if the population has by chance in any given year exceeded the carrying capacity of the environment to support it. In Galápagos the maximum carrying capacity is likely about 2 000 penguins. To invoke carrying capacity, each year you would simply ask if the predicted population exceeded 2 000. If it did, you would set the population to 2 000. If it did not, you would leave it as it was.

This is all you need to understand to build a simple model of the population in a spreadsheet environment.

Converting Conceptual Steps into a Simulation Model

Now let's turn to a spreadsheet program. We will focus on MS Excel® as it is the most widely used such program. Because most spreadsheet programs operate in a similar manner the directions below should suffice for any spreadsheet environment.

Once you have opened MS Excel®, type these data in the cell indicated:

A1 : 7
B1 : El Niño frequency (years)
A2 : 10
B2 : Quasi-extinction threshold
A4 : "Year"
B4 : "Factor"
C4 : "Size"
D4 : "Status"

... such that your spreadsheet looks like this:

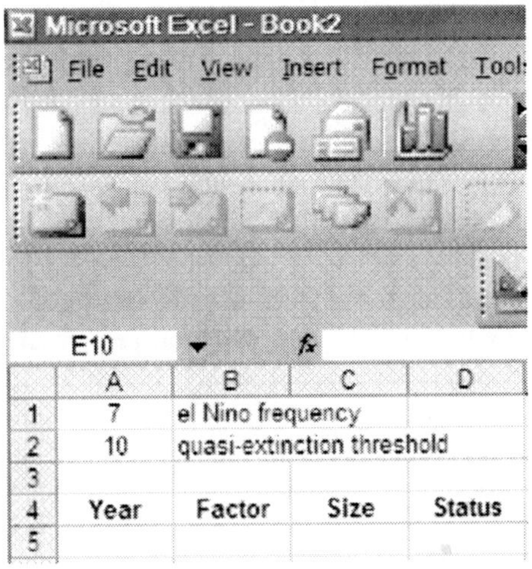

Fig. 9.2

Establish the Sequence of Years

Under Year in cell A5, type 0 for year zero. In cell A6 type "=A5+1." As you will see this gives the cell a value of 1 because A5 contains 0 and $0 + 1 = 1$. Next, copy the contents of cell A6 and paste them into cells A6 through A104. Now you have a full sequence of years from 0 to 99.

Establish the Population Growth Factor for a Given Year

Under Factor in cell B4 enter into B5 the following equation:

$$= IF(RAND() < 1/\$A\$1, 0.31, 1.188)$$

Examine this formula. It does exactly what we described above conceptually, that is, if a uniform random number is drawn that is < 1/El Niño frequency, then we will choose the El Niño population growth factor of 0.31; otherwise the value is greater and we will use the normal year population growth factor of 1.188. This is because the Excel "IF" statement uses the following logic:

$$= IF(\text{logical test, value if true, value if false}).$$

Note, that the $\$A\$1$ means that when you go to copy the formula that cell address will not change. So now copy the formula from B5 and paste it into B6 through B104. Having done so you will see you have a long sequence of population growth rates for El Niño or normal years depending on the random number drawn for any given year.

Project the Penguin Population over the 100-Year Period

Now go to cell C5 right under Size for population size. Enter into C5 the value of 500 – the starting population of penguins. Next go to cell C6 and enter the following formula:

$$= IF(C5 \times B5 > 2000, 2000, IF(C5 \times B5 > \$A\$2, C5 \times B5, 0))$$

This is the formula that will project the population from one year to the next with the appropriate population growth factor and invoking carrying capacity or quasi-extinction if needed. Copy this formula from C6 and paste it into cells C47 through C104. Now you have a sequence of population sizes over 100 years that correspond to the previous year's population in the random draw of an El Niño versus a normal year.

Now let's examine this formula in detail based on the cell addresses and the logic of the IF statement. This is in fact two IF statements working in tandem (a so called "nested" if statement). The formula basically indicates

If the population in cell C5 (the population size the previous year and currently the starting population of 500) multiplied by the population growth factor for this year (residing in B5) is greater than the carrying capacity (2 000) then let the population grow no further by setting it to 2000. If, however, it is less than 2 000 look at it again: if it is greater than the quasi-extinction threshold (indicated in cell $\$A\$2$) then project the population forward by multiplying the previous years population by the appropriate growth factor for the year BUT if this is not the case then the previous population had to be below the quasi-extinction rate so set the current population to zero (that is, extinct).

Determine Whether the Population Projection Fell Below the Quasi-Extinction Level at Some Point over the 100 Years

To do this go to cell D5 under "Status" and enter the following formula:

$$= IF(MIN(C5:C104) = 0, \text{"Extinct"}, \text{"Extant"}).$$

Again, following the logic of the IF statement, we will search through the population values for all 100 years and if the minimum value for any of them is 0 then the population went extinct; if not, it will be considered to have remained extant. We use this "quasi-extinction" level of 10 because if the population were to actually drop to such a level lots of problems would arise, particularly those associated with inbreeding, and it would not likely recover. We could use 25 but 10 is a conservative quasi-extinction level and certainly more appropriate than 0.

Graph Your Population Trajectory

One of the handy things about MS Excel® is that it continually updates any graphs that you make every time you recalculate your spreadsheet. This can give you a real-time sense of the dynamics of your population. To graph the dynamics of your population over time first highlight all the cells for years 0 through 99. Then, **holding down the Ctrl key**, highlight all the cells containing the associated population sizes. Then choose Insert/Chart and choose XY (scatter) and again choose the scatter points connected by lines and click on Finish. Now your population size by year graph will appear and you can place it at any convenient spot.

At this point you should have a spreadsheet that looks much like Figure 9.3.

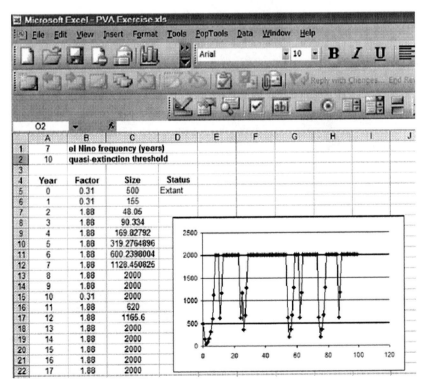

Fig. 9.3

Table 9.1 Tally form for your PVA model simulations. Extinction probability is simply the number of iterations ending in extinction divided by the number of iterations for a given scenario of El Niño frequency.

El Niño frequency (years)	Iterations	Number iterations ending in extinction	Extinction probability
7			
6			
5			
4			
3			

Iterate Your Model

The last step is to iterate your population model to simulate dynamics of the penguin population over time. To do this you simply press F9 and a whole new set of random numbers is generated. Upon doing this the various IF statements also go to work, interpreting the new random values, determining if it is an El Niño year or not, and passing the population from one year to the next based on the rules you used.

To actually perform the PVA for this population, repeatedly press F9 100 times and keep track of the number of number of times the population went extinct in Table 8.1. Repeat this process for values of El Niño frequency for 7, 6, 5, 4, and 3 and enter the corresponding values into Table 9.1. Last, calculate the extinction probability for each scenario, that is, the fraction of iterations in which the population went extinct.

Expected Products

- A fully functioning population-viability model in spreadsheet format
- Estimates of penguin population quasi-extinction rates under El Niño event frequencies of once every 7, 6, 5, 4 and 3 years.
- Responses in a form indicated by your instructor to the Discussion questions below.

Discussion

1 Do you consider the specter of increasing El Niño frequency associated with climate change to be one for concern for the Galapagos penguin? Be specific in terms of changes in your estimated extinction probabilities.
2 What is a reasonable extinction probability threshold to consider? 1%? 5%? 10%? 50%?
3 What is a reasonable quasi-extinction threshold to use? Is 10 penguins reasonable? Consider rerunning the simulations with another and examine the results.

4 PVAs are obviously useful tools for gaining insights about populations. However, what might be some drawbacks to their use for making management recommendations?

5 An outcome from this analysis might be to maintain normal El Niño event frequencies. Explore the practical, political and perhaps ethical issues associated with accomplishing these objectives.

Making It Happen

Application of PVAs is often difficult because of the meager amount of information available on the natural history of most species, especially those threatened with extinction. PVAs should be regarded primarily as a tool for guiding research, management, and policy, rather than for making definitive statements about the fate of populations. Nevertheless, without PVAs, many insights about threatened populations are unavailable. In addition to the formal PVA analysis, the PVA process is often a valuable exercise in gathering all the information available for a species and synthesizing it, thereby forcing the researcher to become intimately familiar with a species' biology and more aware of the constraints a species is facing. Consider undertaking a new PVA for a species of concern to your local authorities. We have listed a number of references and resources below for this purpose. Whether you succeed in pulling all the necessary pieces together is not critical – simply attempting to undertake the PVA process and assembling all the information available on the species' biology can be very insightful and could generate a valuable document for local authorities.

Further Resources

For those intrigued about PVA and its applications we strongly recommend a book on the topic by Beissinger and McCullough (2002). Also, state-of-the-art PVA software is available for free in the form of Vortex (www.vortex9.org/vortex.html) and for fee in the form of the RAMAS family of software (www.ramas.com/).

CHAPTER 10

Habitat Loss and Fragmentation: Ecological Traps, Connectivity, and Issues of Scale

James P. Gibbs

Sit at the border between a forest and a field for awhile and you could list many profound differences between these ecosystems. Differences in vegetation and microclimate will be especially obvious to you and they are also apparent to animals that need to select appropriate habitat. Some species prefer these "edge habitats," perhaps because they need different resources from different environments, such as fields for foraging and forests for cover. Other species seem to avoid edges; for various reasons they need core habitat that is some distance away from the nearest edge. These edge-sensitive species do not fare well in landscapes fragmented by human activities because as patches of natural habitat become smaller and smaller, the core portion of these patches gets smaller and smaller too. Worse yet, as the area of a patch shrinks, the area of its core shrinks even faster because the edge zone occupies a relatively larger portion of the patch. If the fitness of individuals is lower in edge habitats this could affect the fate of a whole population. Specifically, populations occupying mostly edge habitats may show a decreasing pattern of abundance over time while those in core habitats may be stable or increasing.

Because extensive fragmentation is a recent phenomenon in the evolutionary history of many edge-sensitive species, organisms often lack the ability to discriminate between edge and core habitats. Consequently, many individuals can find themselves in an "ecological trap," that is, living in edge habitats that appear suitable, but in fact are not. Curiously, the extent of edge effects is strongly affected not just by the amount of habitat loss but also by the scale at which it occurs. In other words, two identical amounts of forest loss (e.g. 50% forest loss) can have dramatically different effects on the biota if the loss occurs in 1 hectare as against 10 hectare blocks even though the aggregate loss is the same.

This exercise explores these concepts. You will use maps provided to visualize the habitat fragmentation process and to estimate how various levels of habitat fragmentation affect the proportions of edge versus core habitats present in the landscape as well as the connectivity of the landscape. This exercise focuses on forest fragmentation and an edge-sensitive plant – white trillium – as well as an open-habitat avoiding, mobile turtle – wood turtle – but keep in mind that it can be extended to any extensive type of ecosystem subjected to human disturbance that may occur where you live (e.g. grasslands, woodlands, or even wetlands).

Objectives

To examine how varying levels of habitat loss in concert with the spatial scale at which fragmentation occurs affect:

- The relative proportions of edge versus "core" habitats present in a landscape
- Populations of edge-avoiding, edge-preferring, and edge-insensitive species
- Connectivity and efficiency of movement through the landscape.

Procedures

Landscape Composition of Forest Interior, Forest Edge, and Non-Forest Habitat

For this exercise, you are concerned with a rectangular landscape of 360 hectares (1.8 km East–West by 2.0 km North–South) whose native vegetation cover is mature forest. You are provided two sets of maps with similar sequences of forest loss (100% to 0%) but in one case the forest was lost in 1 hectare blocks (Figure 10.1) and in the other case in 9 hectare blocks (Figure 10.2). (Note that to add some realism to this exercise these maps were made with a geographical information system in which blocks were randomly chosen to be forested or not. Consequently, the actual amount of forest remaining in the landscape is very close to but not precisely the same as the specified level.) Last, for this part of the exercise **ignore the left- and right-most columns** (we will use these in the second part of the exercise) so, in other words, calculate habitat composition on the basis of a 1.8 × 2.0 km landscape (18 habitat blocks by 20 habitat blocks).

The first step in this exercise is to simply describe changes in the composition of the landscape. To do so, tally the hectares of each landscape that is in forest interior, forest edge, and non-forest habitat. Remember to exclude from your tallies the 1-hectare-wide columns on the east and west sides (we will use these next in the connectivity exercise). Table 10.1 is provided to help you organize your data. There is a total of 360 hectares to tally for each scenario (that is, 18 hectares along the bottom by 20 hectares up the sides).

Visualize changing composition of the landscape by making plots of overall forest loss (%) (x-axis) versus:

1 area (hectares) of forest interior habitat present
2 area (hectares) of forest edge present
3 percentage of the remaining forest that is forest interior habitat.

Construct these plots for both the 1-ha-scale loss scenario and the 9-ha-scale loss scenario (best to overlay these curves).

Populations

You will focus on the implications of these landscape changes for population persistence of a forest understory herb, western white trillium (also known as western wake robin, *Trillium ovatum*), a member of the Trilliaceae family, that occurs in rich forest in western United North America. Studies by Jules (1998) indicate that trillium populations within 65 m of forest-clearcut edges reproduce poorly for decades. The cause of this severe "edge effect" is apparently mice, which are more abundant near edges and depredate trillium seeds (Tallmon et al. 2003). Trilliums far from forest edges are able to reproduce adequately to sustain populations. These edge effects on reproduction reduce the trilliums' fitness, such that (for the purpose of this exercise) populations occupying edge habitats have a slightly negative growth rate ($r = -0.02$), that is, the trilliums cannot quite sustain their numbers when living only in edge habitats. In forest interiors, however, population growth rates are positive ($r = 0.02$).

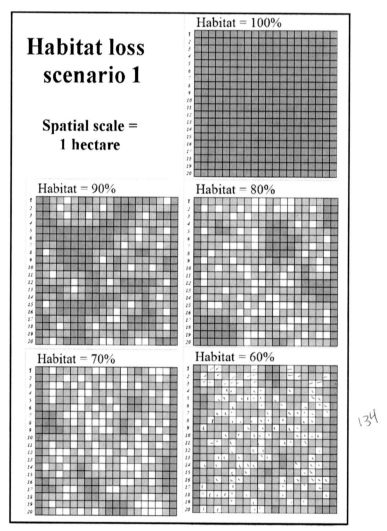

Fig. 10.1 Forest loss sequence (0% to 100% loss) with forest removal occurring at a spatial scale of 1 ha. Dark gray blocks are forest interior, light gray areas are forest edge, and white areas are non-forest. Each block = 1 hectare.

Next, predict the fate of this plant in each landscape. To determine population growth rate for each landscape, calculate the weighted average of growth rates across edge and core habitats. The weighted average is calculated because the population growth rate should reflect overall habitat quality as a function of the changing mixture of edge and core habitats across the landscape. To calculate the weighted average population growth rate, use the % of remaining forest in edge versus interior (Table 10.1) as "weights." Do this simply by multiplying the population growth rate for a particular habitat type ($r_{core} = 0.02$, $r_{edge} = -0.02$) by the % of that forest type present on the landscape, adding the two values together, and dividing by the sum of the weights, that is, 100. For example, if habitats are 90% intact, and core habitats comprise 30% of the intact habitat and edge habitats comprise the remaining 60%, then the average population growth rate will equal $((0.02 \times 30) + (-0.02 \times 60))/100$.

From your inputs in Table 10.1, construct a graph to describe the fate of the trillium in these landscapes. More specifically, graph the estimated population growth rate of the trillium population (y-axis) in relation to the % of forest remaining in the

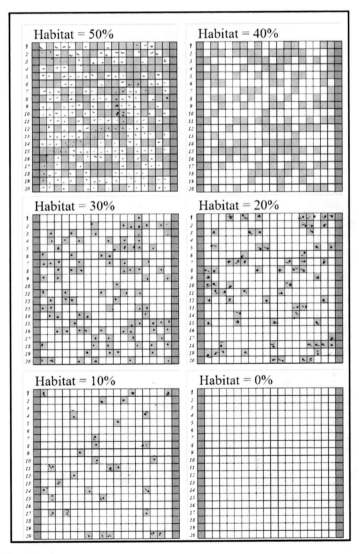

Fig. 10.1 (*Continued*)

landscape (*x*-axis). Do this for both spatial scales of habitat loss. Remember that positive population growth rates mean that the population will increase whereas negative population growth rates indicate that the population will decrease. Note that this will only reflect growth rates of trillium in the forest that remains.

Habitat Connectivity for a Dispersing Animal

For this part of the exercise, you are concerned with the conservation of a mobile animal – a woodland turtle such as the wood turtle (*Glyptemys insculpta*). Consider for this exercise that the forest is the primary habitat but that it gets replaced by suburban development that the turtles avoid. A turtle must travel between two critical habitats (nesting area and home range) annually to survive and reproduce. Your job is to determine how much loss and fragmentation of the intervening habitat the turtles will tolerate before they are unable to successfully complete

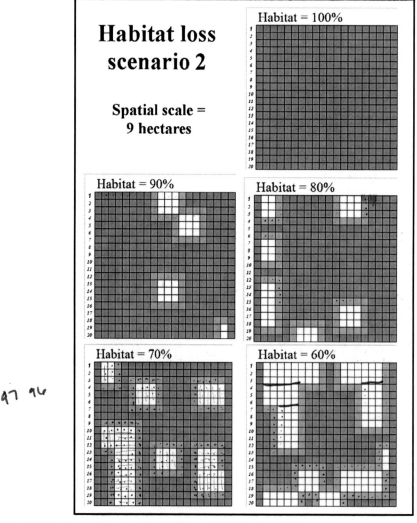

170 97 96

Fig. 10.2 Forest loss sequence (0% to 100% loss) with forest removal occurring at a spatial scale of 9 ha. Dark gray blocks are forest interior, light gray areas are forest edge, and white areas are non-forest. Each block = 1 hectare.

their migration circuit. You also seek to determine how the energetic burden of crossing the fragmented habitat changes as habitat loss levels increase.

You will complete the exercise using the maps in Figure 10.1. For each level of habitat loss, start along the left side of the map and try to pilot a turtle across the landscape *along the shortest route possible* between its starting point in the upland habitat and anywhere on the right side of the map. *You can move horizontally, vertically, or diagonally.* Conduct five trials for each level of habitat loss (that is, 0%, 10%, 20%, ..., 100%). To do so we suggest starting on the left side at point 1, 5, 10, 15, and 20. You can move through either type of forest habitat (edge or interior). Keep track of the number of steps (1-ha blocks, horizontal, vertical, or diagonal) required to move from your starting point to any point on the far right side. Don't worry if your paths cross – you are just trying to get across the landscape. At the end of each attempt, record the total moves required to make the most efficient crossing of the landscape, or enter a dash if you simply could not make it across.

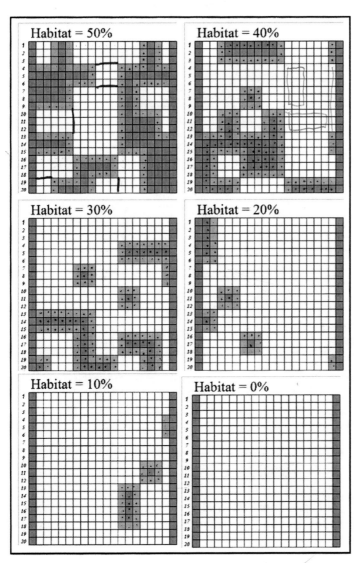

Fig. 10.2 (*Continued*)

Table 10.1 Form for tallying landscape composition and calculating trillium population growth rate.

Forest loss at spatial scale = 1 ha

Forest %	Forest interior (ha)	Forest edge (ha)	Non-forest (ha)	% Forest in interior	% Forest in edge	Trillium r
100	360	0	0	100.0	0.0	0.020
90	141	183	36	39%	51%	-.003
80						
70						
60						
50						
40						
30						
20						
10						
0	0	0	360	0.0	0.0	0.000

(handwritten annotations: "black" above Forest interior; "grey" above Forest edge; "white" above Non-forest; "If positive: increasing; If negative: decreasing →" next to Trillium r)

Forest loss at spatial scale = 9 ha

Forest %	Forest interior (ha)	Forest edge (ha)	Non-forest (ha)	% Forest in interior	% Forest in edge	Trillium r
100	360	0	0	100.0	0.0	0.020
90						
80						
70						
60						
50						
40						
30						
20						
10						
0	0	0	360	0.0	0.0	0.000

Once you have finished running the trials, calculate the efficiency of crossing the landscape in relation to the amount of forest remaining. The relative effort of crossing can be indexed as:

$$\text{Relative Effort} = Moves_{observed}/Moves_{theoretical}$$

where $Moves_{observed}$ = the number of moves required to cross any given landscape divided by the shortest crossing, that, is 18 (recall that the forest tract is 18 1-hectare blocks wide). Calculate relative effort for each trial and then average for each level of habitat loss.

Last, to make inferences about the relationship between habitat loss and connectivity construct two graphs. The first should relate degree of habitat loss on the horizontal axis (0% to 100%) to the fraction of trials you successfully traversed (no matter what the effort level). The second should relate degree of habitat loss on the horizontal axis (0% to 100%) to the average relative effort for successful traverses on the vertical axis. Connect the points in both graphs with a line. For the relative effort graph note that at high levels of habitat loss you will simply not be able to cross the landscape so you cannot plot these points.

Expected Products

- Completed Table 10.1 and with the calculations for trillium population growth
- A single graph (with two lines) of the relationship of forest remaining (%) (x-axis) to (i) forest interior (ha) and (ii) forest edge (ha) (y-axis) at the 1-ha spatial scale of habitat loss
- Similar graphs for the 9-ha spatial scale of habitat loss
- A single graph (with two lines) of the relationship of forest remaining (%) (x-axis) to % forest remaining as interior (y-axis) at the two spatial scales of habitat loss
- A single graph (with two lines) of the relationship of forest remaining (%) (x-axis) to % forest remaining as edge (y-axis) at the two spatial scales of habitat loss
- A single graph (with two lines) of the relationship of predicted trillium population growth rates (y-axis) to forest remaining (%) (x-axis) at the two spatial scales of habitat loss
- A graph of the relationship of forest remaining (%) (x-axis) to the fraction of times the turtle could cross the landscape and its crossing efficiency
- Responses in a form indicated by your instructor to the Discussion questions below.

Discussion

1 Was loss of forest interior a simple linear function of overall loss of forest habitat? If not, how would you describe it? What are the implications of this general pattern for land-use policy and planning?

2 Based on the graphs you constructed, what can you conclude about the relationship between habitat fragmentation and the prevalence of edge habitats and hence edge effects on the trillium population?

3 Because any declining population will eventually go extinct, what is the maximum amount of fragmentation that you would recommend for this landscape to retain a persistent population of the trillium over the long-term? Why is the population in decline over much of the fragmentation spectrum, even in situations where there is a considerable amount of habitat still intact?

4 How might the fate of the trillium population change if the difference in reproductive success between edge and core habitats was greater than the values used in this example? How about if the difference was less than the values used?

5 What are the effects of similar levels of habitat loss occurring in different spatial scales for the fates of core habitats and edge-sensitive species? In particular, how might the pattern of fragmentation, independent of the level of fragmentation, be factored into a management plan to minimize edge effects?

6 How would you design a study to assess whether edge effects influence the biology of a particular species? What kind of data do you need? How would you obtain them?

7 What can you conclude about the relationship between habitat loss and fragmentation and the dispersal probabilities in these turtles? Is it a linear and gradual relationship or is it a threshold relationship? Would you recommend a minimum tolerable level of habitat loss for the habitat intervening the upland and wetland areas? If so, what level would it be?

8 For this exercise you effectively enjoyed a "bird's-eye view" of the landscape and made decisions about where to move accordingly. Many organisms migrate on the ground, however, and will only be able to make decisions based on what habitats are directly visible in front of them. How might these different perspectives affect the likelihood of an organism successfully navigating through a fragmented landscape?

Making It Happen

Identify a management plan for a local public forest, perhaps a local, proposed housing development, or even a local road construction project. Look at the plan carefully with an eye toward how the plan might contribute to fragmentation of local habitats and identify any edge-sensitive species likely to be affected. Can you identify opportunities for altering the plan to minimize its fragmentation-producing aspects? Can these changes be implemented while permitting the plan to meet its original objectives? If not, what threshold level of fragmentation of the land, and hence disturbance or development, is tolerable from the viewpoint of the edge-sensitive species affected? How might you get your views incorporated into the planning process?

Further Resources

We recommend Lindenmayer and Fischer (2006) as an excellent overview of habitat fragmentation issues. These recent reviews of the effects of habitat loss and fragmentation on plants and animals are also useful: Roslin (2002), Debinski & Holt (2000), Sih et al. (2000), Smith and Hellmann (2002), Reis and Sisk (2006) and Reis et al. (2006). For more background on the trillium biology and its relation to forest fragmentation see Jules (1998) and Tallmon et al. (2003).

CHAPTER 11

Diagnosing Declining Populations: Assessing Monitoring Data to Better Understand Causes of Rarity in an Endangered Cactus

James P. Gibbs

We need to identify the factors causing the declines in any endangered species before we can enact effective management actions to reverse declines and "recover" a species. A flurry of theoretical prognostications can be applied to shed some light on the problems affecting a declining species but, if time and resources permit, a more productive approach is to systematically monitor a declining species in the field to diagnose what ecological factors underlie the declines. Once analyzed, the monitoring data can then point out some specific directions for further investigations to pin down the causes of imperilment and to identify what can be done about them. Basically this involves applying the scientific method to make a diagnosis. More specifically, Caughley and Sinclair (1994:224) outlined the five steps to follow in order to properly diagnose causes of population decline:

1 Confirm that declines have indeed occurred by comparing current population counts with comparable counts made at an earlier time.
2 Understand the species' natural history so that you can make useful hypotheses about why it might be declining.
3 Given your understanding of the species' natural history, list all reasonable and potential causes for the decline.
4 For each possible agent of decline, measure its current and historical level. Corresponding trends in the levels of a possible decline factor and those of the species' abundance suggest a possible cause–effect relationship that should be tested as a research hypothesis.
5 Proper scientific studies, preferably experimental in nature, should then be conducted to determine whether an agent of decline is causally related to the species' decline and not simply correlated with it.

In this exercise we apply Caughley and Sinclair's (1994) approach to an endangered species of cactus. We use the Chisos Mountains Hedgehog Cactus (*Echinocereus chisoensis* var. *chisoensis*) as a model species to make the scenario realistic (but the situation is hypothetical).

By way of background, this small cactus has cylindrical, mostly single stems up to 20 cm tall. It inhabits low-elevation desert grasslands or sparsely vegetated shrublands on gravelly flats and terraces within the Chihuahuan Desert. It flowers from March–July with fruits maturing from May–August. Very limited vegetative reproduction also occurs. Usually the plants are found in the sparse shade offered by desert shrubs, which function as nurse plants. Nurse plants offer a slightly moister environment for seed germination and seedling establishment as well as protection from herbivores. The cactus' pollinators are unknown, but native bees are suspected to be the primary mechanism of pollen transport between plants. The only known disperser of seeds is the collared peccary (*Tayassu tajacu*). Peccaries eat the fruits fallen below the adult cactus and defecate the seeds elsewhere. Once widespread in the Chihuahuan Desert, the cactus is now known from fewer than 20 sites, all of which have populations of fewer than 500 individuals. It is not known if these populations are linked through gene flow.

The cactus is listed as "threatened" and reasons for the cactus' declining status are not well known. Although somewhat inconspicuous when not in flower, the large, bright pink flowers are a target not only for insects but also for cactus collectors ("poachers"), who prefer the larger individuals (those that flower more frequently) for their collections. Other possible causes of decline include livestock (mainly cattle). Cattle impact the species directly by trampling individuals of all sizes and indirectly by destroying nurse plants and thus hindering cactus regeneration. Collared peccaries also may have declined thereby limiting the capacity of the cactus to disperse to and colonize new areas and hence maintain genetic flow among its populations. Because cactus populations are reduced from their former levels and now occur as small, isolated collections of individuals (< 500), it is possible that inbreeding effects in this normally outbreeding, widespread species are reducing population growth. Such inbreeding effects, if important, would be most visible in reproductive characteristics and more pronounced in smaller than in larger populations. A last factor to consider is pollinators. No data are available on bee populations, but information is available on seed-set, that is, the proportion of flowers produced by cacti that develop into fruits (seed set in cacti below 75% might suggest inadequate pollination).

What are the likely causes of this species decline and what can we do about them?

Objective

- To learn how to systematically investigate the factors responsible for population declines.

Procedures

Field Data

You are provided below with a set of data with which to diagnose causes of declines in an endangered cactus. Table 11.1 presents population monitoring data from

Table 11.1 Monitoring data on cactus numbers and sizes from two time periods.

| | Number of cacti | | | | | |
| | 1988 | | | 1998 | | |
Site	Small	Medium	Large	Small	Medium	Large
1	32	16	8	9	5	2
2	379	186	91	41	20	10
3	42	21	10	23	11	5
4	20	10	5	16	8	4
5	22	11	5	5	2	1
6	1	1	1	3	2	1
7	213	104	51	70	35	18
8	43	21	10	48	24	12
9	13	6	3	30	15	7
10	45	22	11	8	4	2
11	89	44	22	19	10	5
12	26	13	6	12	6	3
13	9	4	2	5	2	1
14	2	1	1	8	4	2
15	101	49	24	18	9	5

Table 11.2 Ecological data from the same sites and times as the monitoring data.

| | Ecological data | | | | % seeds developing |
| | Cattle density | | Peccary density | | |
Site	1988	1998	1988	1998	1998
1	33	173	86	62	66
2	14	51	5	56	94
3	33	49	76	94	80
4	71	137	34	12	90
5	32	195	48	40	10
6	71	78	29	36	9
7	23	69	86	19	97
8	55	78	52	49	75
9	36	24	10	80	98
10	5	27	78	10	92
11	27	74	16	88	98
12	64	148	93	51	50
13	63	146	9	13	17
14	91	22	69	15	12
15	11	54	42	18	92

two time periods and 15 sites. Table 11.2 presents monitoring data for ancillary ecological variables that pertain to the possible changes in the cactus population.

Now let's revisit Caughley and Sinclair's (1994:224) first four steps to try to determine what is happening to this species and why:

1. Confirm that declines have indeed occurred by comparing current population counts with comparable counts made at an earlier time

How will you do this? Here are some pointers. You have monitoring data from the same populations for two time periods. Estimating population change is straight-forward with this kind of data. To what degree have these populations changed? The best way to measure this is to subtract the second counts from the corresponding first counts for each site. If populations have on average gotten larger then the mean difference should be positive. If populations have on average gotten smaller then the mean difference should be negative. Calculate the average net change across populations. If you wish to approach this statistically, use a so-called "paired-sample" t-test (also known as a paired t-test). Most spreadsheets include this standard function (including Microsoft Office Excel® and the freeware equivalent that is part of OpenOffice Suite at: www.openoffice.org/). Make sure to choose the "paired" option and choose the correct columns when making this test. Note that you can test for changes in aggregate numbers as well as in particular size classes. Did populations of this species decline on average between 1988 and 1998?

2. Understand the species' natural history so that you can make useful hypotheses about why it might be declining and 3. Given your understanding of the species' natural history, list all reasonable and potential causes for the decline

What might be the most likely causes of a decline in its populations? List each possible cause and identify what might be the most likely pattern in the data provided if these agents of decline were important. Develop a prediction for how each potential decline factor, if it were important, would cause a change over time in the variables monitored, including particular size classes of cacti.

Here are some general hints to help you test your predictions.

- Poachers are selective upon larger individuals, which produce the showy flowers they seek, and tend to ignore smaller individuals.
- At any given site, assume increases in seed-disperser abundance should facilitate population growth in this rare species, whereas increases in destructive agents should be associated with population declines.
- Last, if this is a small population struggling with inbreeding, then some measure of reproductive success should be correlated with population size.

4. For each possible agent of decline, measure its level where the species is currently and historically. Corresponding trends in the levels of a possible decline factor and those of the species' abundance suggest a possible cause-effect relationship that should be tested as a research hypothesis.

Here you should find a way to analyze the data to draw inferences about the strength of the predictions you made in Steps 2 and 3. If you have some background in statistics you may wish to use a formal statistical approach to generate your conclusions but a graphical approach to analysis (plots of change in one variable against change in another) will suffice to complete the exercise.

Expected Products

- A list of possible causes of decline
- A description of likely patterns in the monitoring data if particular agents of decline were important
- An assessment of whether the species' populations have indeed declined or not
- An analysis of the monitoring data to determine which of the patterns in the monitoring data support which of your predictions
- A discussion on the priorities for further, experimental research that should be conducted
- Responses in a form indicated by your instructor to the Discussion questions below.

Discussion

Recall Caughley and Sinclair's (1994:224) Step 5: **Proper scientific studies, preferably experimental in nature, should then be conducted to determine whether an agent of decline is causally related to the species' decline and not simply correlated with it.** The data you have are purely correlative and hence even under the best of circumstances determining causation from them is problematic. So what sorts of experiments could you develop that would directly test which factors of decline were at work? Remember that you are working with an imperiled species and on a small budget. More specifically, discuss the design of studies that you might undertake to determine the relative contribution of cattle, poaches, peccaries, and small population size to changes in the cactus' populations. For starters, what might carefully planned fence-building accomplish to "exclose" some species but not others? Can you think of why experiments with pollen transferred between population on paint brushes might be useful?

Further Resources

We recommend Caughley's (1994) paper that contrasts approaches used in wildlife science and, particularly, conservation biology to study populations and address their causes of endangerment. Caughley and Sinclair (1994) elaborate on their systematic approach to diagnosing population declines.

CHAPTER 12

Estimating Population Size with Line Transects and DISTANCE

Michael E. Meredith

Visitors to tropical forests are typically disappointed to see very few large animals, because their idea of tropical wildlife is based on images of wide open savannahs. Large animals may indeed be rarer in forests than savannahs, but most of their disappointment is caused by the difficulty of seeing anything in the forest. You might be 50 m from an elephant in the forest, but not see it because of the trees and foliage blocking your view; next time you see an image of herds of animals in wide open savannahs, consider how many are within 50 m of the camera – allowing for zoom lenses, of course.

Differences in the "sightability" of animals between habitats are an issue not just for tourists but also for biologists and managers concerned with populations and population trends. Imagine if you were to conclude, based on elephants counted per hour while walking along transects, that elephant populations had increased over time when in fact deforestation simply made the few remaining elephants easier to count (in fact, populations may well have declined).

Many population survey methods provide relative measures of animal abundance (for example, numbers encountered per hour or per km), but variation in them may be a matter of "sightability" (or, more broadly, "detectability") rather than variation in the actual population. Detectability will vary with the habitat, detection methods, season, and often the skill and experience of the observer.

Adjusting for variation in "sightability", and thereby arriving at a more robust estimate of density, is the essence of "distance sampling." With distance sampling, you don't just count individuals but you also measure the distance between them and the point where you are standing (for point counts) or between them and the line you are walking (for line transects). The small extra effort in the field enables you to reap tremendous improvements in the reliability of density estimates. For example, to return to our elephant example, you may find that in the forest you are seeing elephants at just 20 m away at most, while in the savannah you can see them 200 m away and perhaps beyond. If your forest elephant counts are only 10% of your savannah elephant counts, then elephant densities in forest and savannah might actually be equal! Only by recording distances to animals observed can you make the required adjustments for bias in the population abundance estimate caused by habitat.

This exercise seeks to help you understand distance sampling methods using a widely available (and free) computer program called DISTANCE, which is the standard tool for this kind of analysis. You will find no shortage of application of distance methods in population survey work, which is a major preoccupation of field biologists and wildlife managers.

Objective

- To understand the concepts behind distance sampling, the operation of the primary supporting software – DISTANCE, and their combined application to field-based population surveys.

Procedures

Concepts Behind Distance Sampling

It is rarely practical to count all the objects of interest in a large park or reserve. In practice we sample, counting the objects in a few small areas (often called **plots** or "quadrats") and calculating the density from the number of objects recorded and the area of the plots. If we have a large sample of randomly-located plots, we will get a good estimate of the average density over the whole area and we can calculate the precision of our estimate (e.g., confidence interval, standard error, or coefficient of variation).

For plants and objects such as nests or dung-piles, demarcating a plot and searching it carefully to count the objects in it works well. But it doesn't work for animals, which tend to flee as soon as you start searching. Transect walks work better.

Instead of randomly placing plots in the area of interest, randomly placed lines or **transects** are used. You move along the transect, recording the animals detected either side. One approach is to decide how far from the line you can be certain of seeing all the animals which are there, and only record animals which are within that distance of the line. This is called a **fixed-width transect** or "strip transect" and is really just a long, thin plot. The problem with this is the width of the strip: if it's too wide, you will not detect all the animals in it, and your estimate will be too low; if it's too narrow, you will have a smaller sample for a given survey effort, and small samples mean less precise estimates.

An alternative is to make the strip very wide, too wide to be sure of detecting all the animals, but to estimate what proportion of animals we do detect. The key to this is the distance of the animals from the transect line. We assume that we see all the animals on the line, and that the proportion detected decreases further away from the line. This is the concept behind **distance sampling**.

A look at a small data set

During transect surveys in Batang Ai National Park in Sarawak, Malaysia, in 1992, we saw 31 groups of muntjac (barking deer). The perpendicular distances from the transect line to the groups of animals when they were first spotted

Table 12.1

Distance from rentis:	0–4 m	5–9 m	10–14 m	15–19 m	20–24 m	25–30 m
Number of groups:						

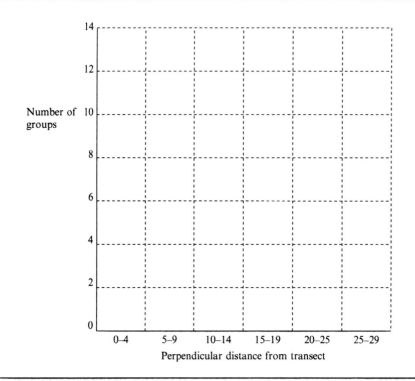

were: 5, 25, 4, 0, 0, 0, 2, 6, 4, 13, 8, 6, 5, 5, 8, 0, 15, 6, 20, 10, 4, 2, 6, 4, 8, 18, 6, 4, 1, 5, 5 m.

Now count the number of groups for different distances from the transect and fill in Table 12.1. Then plot a histogram with the results.

As you can see from the histogram, the number of groups of animals we saw gets fewer farther from the transect. We assume that we see all the animals which are on the transect, and only a small proportion of those further away, as shown in Figure 12.1, which should look rather like your histogram.

The curve in the figure is the **detection function**, symbolized by $g(x)$. The crux of distance sampling is to find the equation for $g(x)$ which best fits the data. The DISTANCE software package will do this for us.

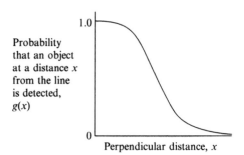

Fig. 12.1

Once we have the detection function, we can proceed in two ways:

1 We can calculate the **effective strip width**, ESW, such that the number of animals detected **outside** the ESW exactly equals the number of animals missed **inside** the ESW (see Figure 12.2A). The calculation is then similar to that for fixed-width surveys, with the area surveyed being $A = 2 \times ESW \times L$, where L is the length of the transect. ("Effective strip width" is a bit of a misnomer: it's really the "effective strip *half*-width".)

2 If W denotes the maximum distance from the line that we recorded animals, as the strip (half-) width, and calculate p, the probability of observing an animal which is present inside that strip (see Figure 12.2B). The actual number of animals in the strip is $N = n / p$, where n is the number of animals seen, and we use N to calculate the density as for a fixed-width strip of area $A = 2 \times W \times L$.

The two approaches are equivalent, since $ESW = p \times W$. This said, the ESW concept may be easier to use, but p is analogous to detection probability in PRESENCE and MARK, and is important theoretically. DISTANCE calculates values for both.

We need one further concept. The histogram you drew from the muntjac data shows how many animals were seen at different distances, while $g(x)$ tells us the probability that an object at a distance x from the line is detected. To model $g(x)$, we use the **probability density function** of perpendicular distances, $f(x)$. For line transects (but *not* point counts), $f(x)$ is the same shape as $g(x)$. Note that on the transect line, where $x = 0$, $g(0) = 1$ and $f(0) = 1/ESW$.

Distance Data for Orang Utan Nests

The muntjac data set we looked at above doesn't have enough observations for a full analysis with DISTANCE. Instead, we'll use line transect data for orang utan nests collected in Batang Ai National Park (Figure 12.3) by June Rubis. The data set has the advantage of having a lot of data points and the fact that distances from the transect could be measured accurately as the nests did not run away before June got the tape measure out.

Like most great apes, orang utan build a simple nest of broken branches to sleep at night, and sometimes build a day-time nest too if they take a siesta. If we assume that there is a simple relationship between the number of nests and the number of orang utan in our study area, we can use the number of nests to compare populations between sites or to detect declines or increases in the population over time.

Download the data file "Bg Ai Nests distances.csv" from this book's website and store it in a folder on your hard disk.

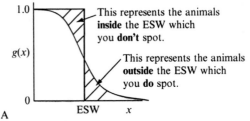

Fig. 12.2 A and B

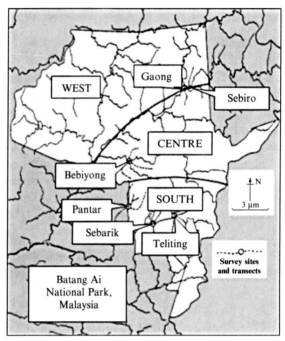

Fig. 12.3 Map of Batang Ai National Park. Data on orang utan nests used in this exercise were collected here by June Rubis.

Open the file in MS Excel® (or other spreadsheet program) and check what it contains.

June worked from 6 sites in the park, with 2 or 3 transects leading out from each site. This is shown in the Transect column. "Length (km)" is the length of the transect: all the transects were 2 km long.

The park was divided into 3 zones, because local people reported seeing lots of orang utan in the south of the park, and fewer in the rest of the park. Surveys in 1992 sighted more animals in the South zone, few animals but many nests in the Centre, and hardly any animals or nests in the West (see map).

Getting Started with DISTANCE

Go to the DISTANCE web site (http://www.ruwpa.st-and.ac.uk/distance/), and fill in the registration form. Then download the installer for the latest version of the software, currently DISTANCE 5.0 release 2 in "d50setup.exe". As with any .exe file, it's wise to download it, run a virus check with an up-to-date virus scanner, and create a Windows Restore Point (go to Start > Programs > Accessories > System Tools > System Restore) before running the setup program. Close all programs (except File Manager) as you may have to restart the computer during installation, then run d50setup.exe. Follow the on-screen instructions. Finally, check the DISTANCE Support Page for any additional patches that may be needed.

If you use the results from DISTANCE in any report or paper, please cite it as:

Thomas, L., Laake, J.L., Strindberg, S., Marques, F.F.C., Buckland, S.T., Borchers, D.L., Anderson, D.R., Burnham, K.P., Hedley, S.L., Pollard, J.H., Bishop, J.R.B. and Marques, T.A. 2006. Distance 5.0. Release 2. Research Unit for Wildlife Population Assessment, University of St. Andrews, UK. http://www.ruwpa.st-and.ac.uk/distance/

Importing the Data into DISTANCE

Open the file "Bg Ai Nests distances.csv" and check that the file is sorted so that all the data for each transect are together and all the transects for each zone are together. (If necessary, use Data > Sort... and sort by Zone first, then by Transect.)

In this case, nests were recorded on all of the transects. In some surveys, there may be transects where you saw no animals: in that case the file MUST contain a row with the name and region for the transect, with the perpendicular distance column left blank.

Now use File > Save As... and save the file as a "Text (Tab delimited) (*.txt)" file, "Bg Ai Nests distances.txt". A box will pop up about "...features that are not compatible with Text...": click Yes. Once that's done you can close the file without saving it.

Start DISTANCE, go to File > New project... and select the folder where you want to save your project – DISTANCE will start off with the Samples folder, but that probably isn't the best place to save your own projects. Name the project something like "BgAi Nests". DISTANCE saves all your work as you go along, you do not need to save frequently. (On the other hand, if you screw up and do **not** want to save your work, don't just close DISTANCE but select File > Revert to Backup Copy and DISTANCE will use the backup copy created when you started the session.)

If you check the target folder now, you will see three new items, files named "BgAi Nests.dst" and "BgAi Nests.ldb" and a folder named "BgAi Nests.dat". The Data Import Wizard will start immediately.

Step 1 : Select "Analyze a survey that has been completed" and click Next.
Step 2 : This is an information page; read through it then click on Next.
Step 3 : The nest survey was a Line transect survey with a Single observer, where Perpendicular distances were measured to Single objects.
Step 4 : The distances from the transect are in metres, the transect length is in kilometers and we want the results for nest density per square kilometer.
Step 5 : We don't need to use any multipliers: all these check boxes should be blank.
Step 6 : Select Proceed to Data Import Wizard and click on Finish.

DISTANCE will create the necessary file structure, and then the Data Import Wizard will start.

Step 1 : This is an information page; check through it then click Next.
Step 2 : In the "File containing data to import" box, browse to the file "BgAi Nests distances.txt" and click Open.
DISTANCE uses a hierarchical structure with

→ a Global layer = study area or topic (in our case Batang Ai nests), containing...
  → Regions (our 3 Zones, West, Centre and South), containing...
    → Transects, containing...
      → Observations.

Step 3: The "Lowest data layer" in our data file is Observation and the "Highest data layer" is Region. We want to Add All New Records under the first record in the data layer and to Create One New Record for each line of the import file.
Step 4 : In our file, the Delimiter is Tab, and we **do not want to import the first row**, as that is just the column titles.

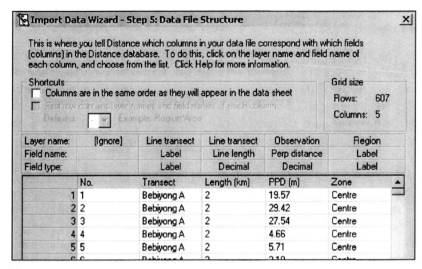

Fig. 12.4

Step 5 : In our case, the columns are not in the same order as in DISTANCE, so we need to tell DISTANCE what each column means (see Figure 12.4 above).

Step 6 : Select Overwrite Existing Data and click Finish!

DISTANCE will take a little while to read your data file and get the data into its own database format, then the Data browser will appear (see Figure 12.5).

Click on Observations **(1)** in the left-hand window to display all the data and scroll down to check the contents. You will notice that the transects all have 2 km as the length, which is correct, but area for each Region is 0 **(2)**. This will cause problems later on when we need to calculate densities for each area, so we'll enter that data now. (We could have included a column in the .txt file with zone areas, but since only 3 numbers are involved it is easier to do it like this, directly in DISTANCE.)

Click on Region in the "Data layers" panel on the left to see all three Regions conveniently. Double click on the first 0, next to Centre. You may see a box pop up with the title "Cannot edit data" explaining that the data sheet is locked. Click

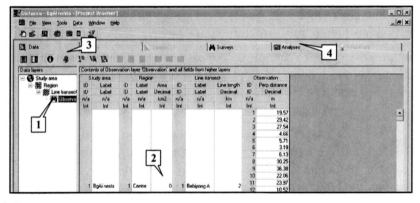

Fig. 12.5

on the padlock icon **(3)** to unlock the data sheet. Double-click on each of the 3 zeros in the Area column and enter the Areas as follows:

- Centre: 74
- South: 56
- West: 128.

Now click on the padlock icon **(3)** to lock the data sheet and prevent any changes being made accidentally.

A First Run Through an Analysis

We'll first run through a complete analysis just with the default settings, to get an overview of how it works.

Click on the "Analyses" tab **(4)** to open the Analysis Browser window (Figure 12.6). You may have to drag the divider between the windows **(5)** to the right to see the full names. DISTANCE has already created a "New Analysis" with the default settings, and we can run that by clicking on the "runner" icon **(6)**. The analysis will take a minute or so to run, during which time your screen may go black – don't panic! When it's finished, the grey bullet on the far left will turn to an orange color and some numbers will appear in the columns in the right-hand panel of the browser (Table 12.2).

That all looks reasonable so far. Let's have a look at the details of the analysis.

Double-click the orange bullet on the left to go to the analysis window. The analysis window opens with the Log tab, which is orange, like the bullet you just

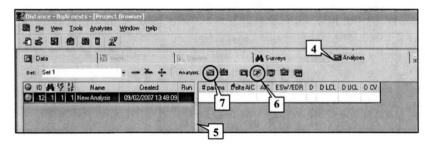

Fig. 12.6

Table 12.2 Computer analysis results displayed in DISTANCE.

# params	1	These are used to compare analyses, so we'll discuss
DeltaAIC	0.00	them later, when we have more than one analysis.
AIC	4179.28	
ESW/EDR	19.02	Effective Strip Width (or Effective Detection Radius for point transects)
D	531.792	Estimate of the density of animals (or, in our case, nests)
D LCL	420.135	Lower and upper 95% confidence limits of D
D UCL	673.123	
D CV	0.112	Coefficient of Variation for D

clicked. That's because the analysis produced warnings, but in this case it's only a warning that we have too many points for all to be included on the q-q plot (more on the q-q plot later). Nothing here to worry about.

Now select the Results tab. Click the Next and Back buttons at the top right of the window to move between the various results pages.

The first page has general information about the default analysis. Some of this is self-explanatory, some items we'll come back to later. The bottom half of the page has a Glossary of terms and symbols which you may find useful.

On the second page, check the 4 numbers top left: effort = total length of the transects (15×2 km); # samples = number of transects (15); width = in this case, the maximum perpendicular distance observed (51 m); # observations = number of observations (duuh!) (607).

Flip through the pages until you find a graph that looks like Figure 12.7.

The histogram (blue in DISTANCE) shows the distribution of the observations. It shows that the probability of detection is highest at distances smaller than 10 m, with that probability decreasing as the distance gets larger. There is a clear "shoulder" to the histogram: the observer seems to have detected all nests out to 10 m and then a decreasing proportion beyond that. The curve (red in DISTANCE) is the fitted detection function, in this case a "normal" or Gaussian curve – or the right-hand half of a normal curve, to be exact – adjusted to fit the observed frequencies (the histogram) as closely as possible. If you go back to the first page, you will see under the heading "Estimators" the entry "Key: Half-normal"; this is the default for a new project. DISTANCE plots three graphs like this with different widths for the histogram bars, and these sometimes show up problems. For example, the second graph has two tall bars and two short bars on the left: if the tall bars corresponded to 5 m and 10 m we might suspect that "heaping" was going on, i.e., the observer was actually estimating distances to the nearest 5 m. All three look okay in this case.

The fit *looks* pretty good, now let's look at the various measures of goodness-of-fit that DISTANCE supplies. You'll see all these if you flick back and forth among the pages of the Results tab:

- chi-squared tests: each of the histograms is followed by a chi-squared table, with actual and expected values for each histogram interval, with the chi-squared

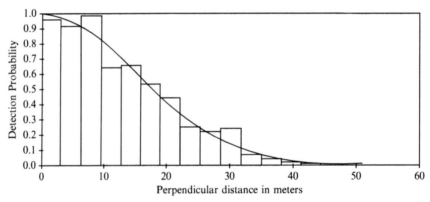

Fig. 12.7 Histogram of perpendicular distances of the observations from the transect.

statistic[1] calculated from the difference. For the first histogram, the probability of getting a greater chi-squared value if the fitted curve is the correct model is fairly low, ie. our data are rather a long way from the modelled values. Not a very good fit. But a point to note is that some of the expected values are pretty low: we should not be doing a chi-squared test with expected values less than 4, and we should be pooling the values beyond a distance of 40 m into a single group for this calculation. In fact we'll adopt a different solution to this problem, as we'll discuss below.

- q–q plot (Figure 12.8): this compares the actual observations (red dots in DISTANCE) with the expected values predicted by the detection function (blue line). If the fit was perfect, all the red dots would lie exactly on the blue line.

- the Kolmogorov–Smirnov and Cramér–von Mises tests are measures of how far the red dots are from the blue line, K–S using the biggest difference and C–vM taking a weighted average. Both then see how likely such a distance is if in fact the model is correct. A high likelihood (close to 1) means a good fit. The cosine-weighted C–vM test gives greater weight to the fit nearest to the transect centre line, which for us is the most important part of the curve (our density estimate depends on where the curve cuts the distance=0 axis). In this case, the K–S value is just about acceptable, and the C–vM results are okay.

The estimates of densities (and total number of nests) are calculated from two elements: the encounter rate (n/L) and the effective strip width (ESW).

Go on to the Estimation Summary – Encounter rates page. This gives the encounter rate (n/L), which is the number of nests per km of transect, plus its confidence interval.

n/L: number of objects (nests) observed per unit length (the encounter rate), $607/30 = 20.233$ nests per km.

Now go on to the Estimation Summary – Detection probability page. This begins with some information on the fit of the Half-normal/Cosine model used to compare models: number of parameters, loglikelihood (LnL), AIC, and two alternatives to AIC

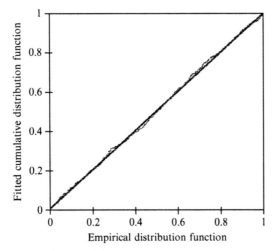

Fig. 12.8 A q-q plot measuring model fit.

---

[1] The chi-squared statistic is named from the Greek letter chi (χ). For each bar of the histogram, calculate $(\text{Observed} - \text{Expected})^2/\text{Expected}$ then add up the results for all the bars.

which we'll discuss when we've run a few models to compare. You should recognize $f(0)$ and p from the theory of distance measurements. The figure we want here is:

ESW: the effective strip width, we'd see the same number of nests if we'd spotted *all* the nests within 19.024 m of the line, and none beyond.

The last page is Estimation Summary – Density & Abundance,

D: the density of nests, 532 per sq km.

N: the number of nests in the whole park, based on the density x the areas we put in for the Regions, 137,200.

DISTANCE also estimates variances: you will see that the % CV for the ESW is quite low, and most of the variance in the density (and number of nests) is due to variance in the encounter rate. We'll return to that point later to see if we can reduce it.

Before we try fitting different models, let's go back to the problem we saw with the chi-squared test.

Truncating the Data Sset

If you look back at the histogram-plus-fitted-curve shown above, you'll see there are just a few observations beyond 40 m. These we saw caused a problem with the chi-squared test, but that is really only a symptom of a bigger problem: fitting the curve to these few outlying points can distort the results which we are really interested in, which is the area close to the line. The solution is to be brutal and discard these distant observations. In fact, looking at the chi-square table, you'll see there are only a few observations beyond 30 m, so we'll truncate at 30 m.

Select View > Analysis components from the pull-down menus to open the Analysis Components window, which has two pages: click on the first button on the toolbar (with the funnel icon) to see Data Filters.

The Default Data Filter should be highlighted. Click on the yellow star (third icon on the lowest toolbar) to create a new data filter. Now click on the Properties button (the fifth on the toolbar, with a hand icon) to open the Data Filter Properties Box. Select the Truncation tab (see Figure 12.9).

Select the "Discard all observations beyond" button and enter 30 in the field. Change the name of the filter to something short but informative, such as Trunc30, then click OK.

Now go back to the Project Browser (use the View menu) and the Analyses tab. Look for the New Analysis button on the toolbar (7 in Figure 12.6) and click to create another analysis. Double-click on the grey bullet next to the new analysis.

The Inputs tab of the new analysis has several panels. At the bottom is the Model definition with the default model. Next is the Data Filters panel with the old default filter and the new filter we just created. Click on the Trunc30 filter to highlight it. Skip the Surveys panel as we only have one survey. At the top change the name to something informative such as "Trunc30 + default model".

Then click on Run and wait... Ignore the warning about the q-q plot and look at the Results tab.

Check first the number of observations: 580 instead of 607, so we have discarded 27 data points. The curve still looks okay, and although the K–S and C–vM and the first chi-square tests are still not brilliant, the last two chi-square tests look more reasonable.

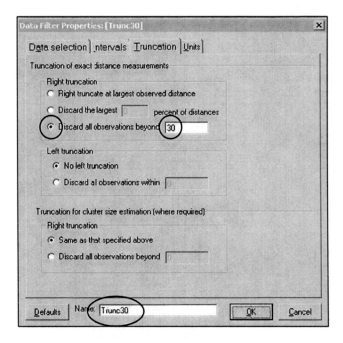

Fig. 12.9

In the Results Browser, we can see that the ESW is about 30 cm less than for the untruncated data, and the density is somewhat lower. Note that we cannot compare the AIC for these two, as they are based on analysis of different data sets.

Now let's go on to compare different models.

Constructing and Comparing Models

DISTANCE offers a range of models with different mathematical equations for the detection function – the red line in the histogram diagrams in DISTANCE. Look at the graphs in Figure 12.10 to get an idea of what these are like. Each model is based on one of three key functions (uniform, hazard rate or half-normal), each of which has a fairly flat area near the transect line and then drops off.[2] To obtain a better fit, DISTANCE tries adjusting the curve by adding in other terms, either cosines or simple polynomial terms (of the form $y = a + bx + cx^2 + dx^3 + \ldots$) or Hermite polynomials, the last being similar to a polynomial but more suitable for adjusting the half-normal curve. The number of extra terms to use is decided automatically; in the example we just used, adding terms didn't improve the fit, so just the plain half-normal curve was used. (Look at the second page in the Results tab and you'll see that DISTANCE tried Model 1 with no adjustments and Model 2 with 1 adjustment and found that Model 1 was better.)

Four combinations are recommended:

- uniform + cosine (also known as the Fourier series)
- hazard rate + polynomial
- half-normal + cosine (the default option we used above)
- half-normal + Hermite polynomial.

[2] The software can also fit a fourth key function, the negative exponential, but this does not have a flatish section at the left end. That means it gives very imprecise estimates *even if* it is a good fit to the data, and it is not recommended.

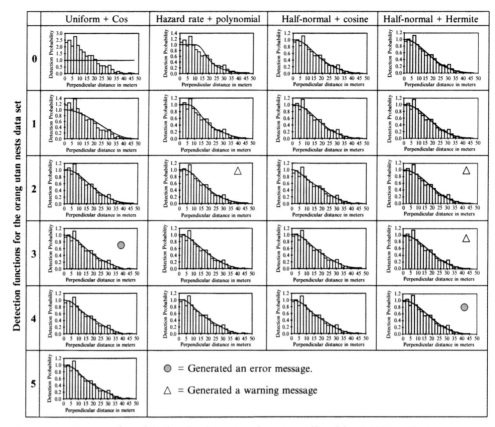

Fig. 12.10 Range of models for the detection function offered by DISTANCE.

Go back to the Analysis components window, but now we want the Model Definitions page. The Default Model Definition should be highlighted. Now open the Model Definition Properties box (see Figure 12.11), go to the Detection Function tab and press Models.

This model is already a half-normal key with cosine adjustment; just change the name at the bottom to something like "Hnorm + cos". While we're here, press the Diagnostics button and change "Maximum num points in qq plots" to something bigger than 600, say 700. Now click on OK. A box will pop up telling you that you'll have to rerun an analysis if you change the model definition – that's not a problem, so click Yes.

Click three times on the yellow star (third icon on the lowest toolbar) to create three new model definitions.

Open the properties box for the second Model Definition and change this to a *Uniform* key function with a *Cosine* series expansion, and change the name at the bottom to something like "Uni + cos". In the same way, change the other two Model Definitions to "Haz + poly" and "Hnorm + Herm".

Now go back to the Project Browser and the Analyses tab. Highlight the "Trunc30 + default model" analysis and click three times on the New Analysis button to create three more analyses.

Double-click on the grey bullet of the "Trunc30 + default model" analysis and go to the Inputs tab. At the bottom of the page you'll see the four models definitions, with the "Hnorm + cos" model highlighted – that's the default model that we renamed. Check that Trunc30 is highlighted in the Data filter panel, then change

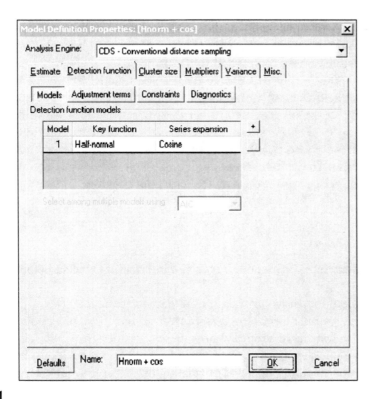

Fig. 12.11

the name of the analysis to, say, "Trunc30 + Hnorm + cos," to indicate which model it uses.

Open another new analysis, but now in the Model definition panel at the bottom click on "Uni + cos" to highlight it, then change the name of the analysis to "Trunc30 + Uni + cos". Do the same with the remaining two analyses, changing the Model Definitions so that we have one analysis for each model.

Run all the analyses with grey buttons: highlight all of them with shift+click or ctrl+click, then click the Run button.

Let's begin with the results in the Analysis browser, ignoring for the moment the first analysis with the untruncated data, which you can now delete if you wish (use the Delete Selected Analysis button on the toolbar). The four should be in order with the one with the lowest AIC at the top[3] (if it isn't, click on the heading of the AIC or deltaAIC column). The "Trunc30 + Uni + cos" model is the best, with an AIC of 3822.05, but the two Hnorm models are close behind. In fact, they are identical: the number of parameters (1) tells us that one model is the Hnorm model with no cosine terms and the other is the Hnorm model with no Hermite polynomial terms! The Haz model is more than 2 AIC units behind, so there is less support for this model.

If you don't get the same results as this, check the analysis setup. This is a bit complicated and it's easy to get the different filters, key functions and adjustments mixed up. To check, look at the second page of the results, headed "DetectionFct/ Global/Model Fitting." Near the top, the Width should be 30.000 and the # observations 580 if you have truncated the data at 30 m. Further down is the heading

[3] For a detailed explanation for AIC (Akaike's Information Criterion), see Box 13.1, pp. 120–4.

Model 2 and the next two lines tell you the key being used (Uniform, Half-normal or Hazard Rate) and the type of adjustment (Cosine, Hermite polynomial or Simple polynomial). Check that these correspond to the name of the analysis.

The difference in the density estimates for the two best models is small, about 2% of the value. The coefficient of variation (D CV), however, is still high, though about the same for all the models. Remember that we saw earlier that most of the variance in the density estimate is due to the variance of the encounter rate (n/L) between the different transects, which is the same whatever model is chosen for the detection function. In the next section we'll look at a possible way to reduce this, but first you should take a look at the detailed results for the three "Trunc30 + ..." analyses.

Differences Between Zones

The park was divided into 3 zones on the basis of information we had before doing this survey.

Ideally we should do separate analyses for the data for each zone (i.e., stratifying by zone). However, we only have 34 observations from the West zone, not enough to calculate a detection curve properly. So our stratification will only affect the encounter rate. We saw above that most of the variance (CV) in the density estimate was coming from the encounter rate, and calculating that separately for each stratum might help to reduce the CV.

We need to create new Model Definitions for the stratified analysis. Open the Analysis Components window and select the Model Definition page. Highlight one of the models and click on the New button. Open the Properties box for the new model and select the Estimate tab (see Figure 12.12). Click on "Use layer type:" and make sure that Stratum is selected as the Layer type. Then, under "Quantities to estimate..." check the boxes as shown: the Encounter rate should be calculated separately for each stratum, but the detection function should be the same for all strata, i.e. it should be global. Change the name, putting "strat" at the end, then click OK.

Repeat this with each of the other three models, so that you have a set of four "... + strat" models.

Now go back to the Analysis Browser, create four new Analyses, and set them up with "Trunc30" as the filter and one of the "... + strat" models. Run all four.

When they have run, take a look at the Results Browser. You'll see that stratification makes no difference to the AIC value. That's because AIC only reflects the detection curve fitting process, and the stratified model uses the same detection curve. (This would not be the case if we had asked DISTANCE to calculate separate curves for each zone.)

Double-click on the green button next to the "Trunc30 + Uni + cos + strat" model and go to the Results tab. Compare the results for the stratified analysis with the unstratified results.

Getting Finished

DISTANCE automatically saves each of the Filters, Model Definitions and Analyses and the results as you create them or run them, so there is no need to save your work before closing. If you do **not** want to save the work you have done during the current session, you can select File >Revert to Backup Copy and DISTANCE will use the backup copy created when you started the session.

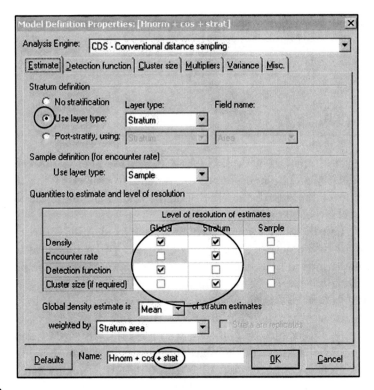

Fig. 12.12

To close DISTANCE, select File | Exit or click on the ☒ button in the top right corner of the main DISTANCE window. When the "Are you sure..." box appears, click on Yes.

Expected Products

- A presentation (in written, verbal, or presentation form as your instructor prefers) of your conclusions about the number of orang utan nests and the density of nests from the DISTANCE output: our analysis has produced several different values and you need to explain which you think is the best.
- A calculation of the population of orang utan in Batang Ai NP, given that one nest is built per orang utan per day, and nests remain visible for between 150 and 280 days.
- Responses in a form indicated by your instructor to the Discussion questions below.

Discussion

1 If you could afford more effort for the next survey, then where would you allocate the effort?
2 If you were asked to design a study to estimate the number of orang utan in a protected area in Borneo, how would you go about it?

3 If you were asked to design a study to estimate the number of some other endangered species, particularly one near where you live, how would you go about it?

Making It Happen

Distance sampling and analysis with DISTANCE are together a powerful tool for estimating populations. The next step is for students to collect data in the field to answer a significant biological question. Exactly what that might be depends hugely on your local circumstances.

For animals, it's best to select a species which is easy to identify and reasonably easy to see, otherwise differences in skills are problematic. You will need at least 60, preferably 80, observations to fit a detection curve; if you want to do separate curves for different habitats, that means 60–80 observations in each habitat.

If a real survey is not practical, a simulated survey can be fun, and highly instructive, as you know what the right answer is. Lay out the transects, then set out randomly-placed objects such as golf tees, wooden pegs, or – in a woodland – flagging tape. Random placement of the objects (using random numbers to determine positions) is itself a good exercise: with two transects, students can set out the objects on one transect, then survey the other. Use two "species" of object, one relatively visible, the other more cryptic and compare the results of the analyses.

Further Resources

DISTANCE itself has good Help pages and an online manual in .pdf format accessible from the Help menu. The standard reference on DISTANCE methods is Buckland et al. (2001), which describes the analysis of several example data sets which are included with DISTANCE once installed on your computer: look in the folder C:\Program Files\Distance 5\Sample Projects. Thomas et al. (2002) give a good overview of distance sampling methods, including many variations on the theme which are not covered in this chapter. Point transects are a common alternative to line transects, especially for bird surveys; Thomas et al. (2002) describe these and Buckland (2006) gives details. Articles on using nest surveys to estimate populations of great apes include Morrogh-Bernard et al. (2003), Walsh and White (2005), and Laing et al. (2003).

CHAPTER 13

Analyzing Camera Trap Data with PRESENCE

Michael E. Meredith

In the last chapter we looked at line transects as a way of estimating the density or population of animals in our study area. Such estimates work well if the animals are spread fairly evenly over the area, so that the true density at each of the transect sites is about the same. If distribution is patchy, so that some transects fall in places with no animals, population estimates will be imprecise – they'll have large standard errors and wide confidence intervals – and it will be difficult to detect trends over time or between sites.

If distribution is patchy, it often makes more sense to monitor the patches rather than monitoring the density or even the overall population. In practical terms, we rephrase our research question from "What is the size of the frog population?" to "What fraction of wetlands is inhabited by a certain frog species?" The results of monitoring specific sites are often more useful in generating testable ideas about the causes of changes and in guiding management decisions than changes in overall population. If frogs disappear from specific wetlands, we can investigate the differences between these and wetlands which are still occupied, and then take appropriate action. In contrast, a fall in overall frog numbers gives few clues as to the reasons.

In principle, determining whether a patch of habitat is occupied or not by a particular species is easy: go and look! If you see or hear the animal, or see identifiable signs such as tracks or dung, you know the species is present. But many species are difficult to detect, and you cannot conclude that they are absent simply because you failed to detect them. In other words, an "absence" can be due to the species actually being absent or the species not being detected due to random chance. The latter is termed a **false absence**. False absences lead to an underestimate of the actual number of sites occupied by the species. Moreover, comparisons between different regions or different survey occasions can be undermined by differences in "detectability" of individuals.

Techniques developed over the last few years allow us to deal with the problem of false absence by estimating the probability of detection from the data. With an appropriate survey design, we can now estimate how many of the sites where animals were not detected are nevertheless occupied, and get a proper estimate of the proportion of sites occupied. As a result, we are increasingly seeing research objectives stated in terms of **occupancy**, e.g. How extensive is the occurrence of a rare herb in a prairie? What proportion of a forest is occupied by tigers? For many cryptic species, it is easier to gather data on presence than to estimate numbers, and a good occupancy study will often be better than an estimate of numbers based on scanty data. But occupancy is only meaningful if the distribution of the species is genuinely patchy. It won't work well for species like hornbills or orang utan, which move over huge areas in search of fruit; it will work for songbirds and gibbons, which defend well-defined territories.

These techniques have only recently been developed partly because the calculations involved could not be done manually. Present-day computers and sophisticated software have now opened up many opportunities for plant and animal monitoring based on repeated presence/absence surveys that can be cheap to deploy. The analysis of such data is somewhat complex, however, and is the focus of this exercise.

Objective

- To explore the concept of "occupancy" and to understand the mechanics of occupancy estimation with the PRESENCE software package.

Procedures

Occupancy of Discrete Habitat Patches

Suppose we had a number of ponds which we thought frogs might use for breeding in the springtime. The calls of breeding frogs are easy to detect in the evening, so we go to our 15 ponds and hear frogs at 6 of them – see column A in Figure 13.1. We might conclude that only 6 out of 15 ponds – less than half – are used by frogs. But we go again and again hear frogs at 6 ponds, but not the same ones (column B).

We are sure that the frogs stay in the same pond throughout the breeding season, so we now know that they are present in 10 of the ponds. Also, even when they are present, we don't always detect them. At the 10 ponds we know have frogs, we only heard calls at 6 on each evening, i.e. 60%.

Can we conclude that there are no frogs at the 5 remaining ponds (grey in the table)? Or could some of them have frogs but we failed to detect them? If the probability of detecting frogs when they are there were $6/10 = 0.6$, and the probability of not detecting them when they are there were $1 - 0.6 = 0.4$, the probability of not detecting frogs on two occasions would be $0.4 \times 0.4 = 0.16$, so it's quite probable that at least one of the remaining 5 ponds could have frogs.

Suppose that one of the remaining five ponds was occupied: that means that we detected frogs in 6 out of 11 ponds, and we'd have to recalculate the detection probability as $6/11 = 0.55$. And then we'd have to recalculate the probability that one of the five ponds is occupied, and then recalculate the detection probability, and then ..., and so on until the recalculations produced the same values (ie. the calculations had 'converged'). That would be a nightmare to do manually, but the PRESENCE package handles it routinely, simultaneously calculating the probability of detecting animals when they are there and the proportion of sites occupied. It can also help us to investigate the effects of covariates such as pond size or air temperature on occupancy or detection probability.

Pond #	A	B
1	✓	✓
2	✗	✗
3	✓	✗
4	✓	✗
5	✗	✓
6	✗	✗
7	✗	✓
8	✓	✗
9	✗	✗
10	✗	✗
11	✓	✓
12	✗	✓
13	✓	✗
14	✗	✗
15	✗	✓

Fig. 13.1 Hypothetical set of presence/absence observations of frogs at 15 ponds.

Occupancy of a Continuous Tract of Habitat

The meaning of occupancy is clear when patches of suitable habitat are well defined and separated from each other; but what does it mean when we are talking about animals in large tracts of forest?

Let's take a small-scale analogy. Suppose that the grey blobs in Figure 13.2 represent patches of lichen on the surface of a rock, and we want to estimate the percentage of the rock covered by lichen. We could try to estimate the area of each blob, add them up, and divide by the area of the rock surface, but this will be quite tedious if the blobs are very irregular in shape. An alternative would be to sample a large number of points at random and see what percentage of these points fall on lichen. Maybe we could take a piece of plastic and punch holes in it at random locations, as indicated by the black dots in the diagram.

Now imagine that the diagram is a map of a forest and the grey blobs are the home ranges of a species of small carnivore, such as a wild cat. Most cats stay within a limited area most of the time, as familiarity with the terrain helps in foraging and avoiding danger; for such species, we can try to estimate the percentage of the habitat that they "occupy." Mapping all the animals' home ranges would mean capturing all the animals, fitting each with a radio transmitter, and then tracking their movements; that would be a huge amount of work, even if we could be sure we'd captured all the animals. As with the blobs of lichen, it's much easier to use sample locations and see what percentage fall within the home range of an animal.

Unlike lichen, however, we can't be sure that the wild cats are absent from a sample location when we fail to detect them. Just as with the frogs in ponds, we need to estimate the detection probability and adjust the occupancy estimate accordingly.

Getting Started with PRESENCE

Download the PRESENCE setup program Setup_presence.zip from: http://www.mbr-pwrc.usgs.gov/software/presence.html, and extract the file

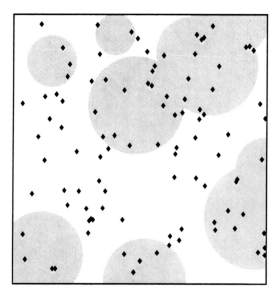

Fig. 13.2 Patches of lichen on the surface of a rock.

Setup_presence.exe. As with any .exe file, it's wise to run a virus check with an up-to-date virus scanner and create a Windows Restore Point before running the setup program. If you use the results from PRESENCE in any report or paper, please cite it as:

MacKenzie, D I; J D Nichols; A J Royle; K H Pollock; L L Bailey; J E Hines 2006 *Occupancy Estimation and Modeling: Inferring Patterns and Dynamics of Species Occurrence.* Elsevier Publishing.

⚠ PRESENCE version 2 is still being actively developed, with modifications being made every few months. The date and time of the version is given at the top of the main window in PRESENCE. The one used in preparing this exercise was 070109.1407. Later versions may behave better, but for the moment there are a few things you should be aware of which are marked with a warning sign. Please check with the book website for any changes affecting this exercise.

⚠ Here's a first warning: Never try to open 2 projects in PRESENCE. Either finish the one you're working on and close PRESENCE (not just the project windows) before starting another project, or start up a new copy of PRESENCE from the Start menu or desktop icon. See below, page 117.

Camera Trap Data for Golden Cats

From December 1997 to October 1999, a major camera trapping survey was carried out at 9 sites in Peninsular Malaysia[1] (Figure 13.3). The main purpose of the survey was to ascertain the status of tigers, but of course the camera traps picked up many other species. In this unit we will look at the camera trap results for golden cats (*Catopuma temminckii*), and see if there is a difference in occupancy for different habitats.

[1] WCS (Wildlife Conservation Society) (2004) Status ecology and conservation of tigers in Peninsular Malaysia. Unpublished report to Malaysian Government. WCS, New York.

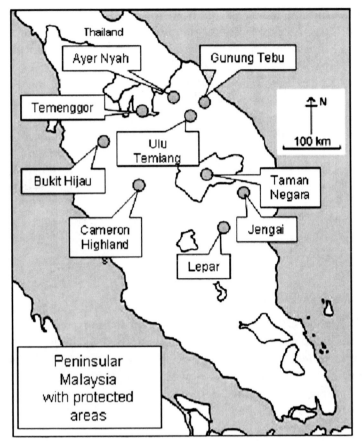

Fig. 13.3 Nine sites in Peninsular Malaysia where wild cat surveys were undertaken in 1997–1999.

The golden cat is a medium-sized cat, standing 50–60 cm high at the shoulder and weighing up to 15 kg. It occurs from Nepal and southern China down to the Malay Peninsula and Sumatra. They are shy forest-dwellers, rarely seen and difficult to study. Automatic, infrared-triggered cameras are one of the few ways to get information on their range and habitat.

Download the data file, Golden cats.xls, from the book website (see page ix) and store it in a folder on your hard disk. Open it in Excel and look at the first worksheet, Raw Data. This gives details of 171 camera trap sites in 9 regions ("zones").

The habitat column indicates for each site if it is logged forest (with the date of logging if known), primary forest (neither logged nor cultivated in the past, with an assessment of quality), or a plantation.

The record of camera trapping has 1 or 0 for each day that the camera was deployed: 1 if a golden cat was photographed, and 0 otherwise. In several cases, the date–time recording system failed, and data from those cameras are not included in the spreadsheet. Cameras were generally operating for a month, but in one case 86 days, and in another case the camera failed after just 1 day. Fortunately PRESENCE does not need equal numbers of observations from each site.

We need to "clean up" the data before passing it to PRESENCE. The raw data are results for consecutive days at each site, so we can't regard them as independent observations. We can improve things by grouping the results into 5-day periods, so that if a golden cat visits a site on consecutive days (as at site T10) it scores one

1 instead of 2. (The same applies if golden cats fail to visit a site on consecutive days.) On the Data for PRESENCE spreadsheet, the results have been grouped into 5-day periods. A couple of sites with less than 5 nights of data have been eliminated from the data set.

We do not have the date of logging for all sites, or an indication of the quality of the primary forest at all sites, so we will simply group sites as "logged", "primary" or "plantation". Once the sites are sorted by habitat, it's soon obvious that no golden cats were recorded at any of the plantation sites. This could be because golden cats never venture into plantations, or because the probability of photographing one is too low. With no data to go on, PRESENCE will not be able to tell you, so we exclude those sites from the analysis. We'll come back to the plantations data in the section below on simulations.

This leaves us with 162 camera sites, split between primary and logged habitats, with detection data for up to 17 occasions (each 5 days). This is in the worksheet Data for PRESENCE within the file, Golden cats.xls.

Importing the Data into PRESENCE

Open PRESENCE and select 'File > New project' from the pull-down menus. Change number of Site Covars to 2. Then click on the 'Input Data Form' button.

The data input form looks like a spreadsheet with 20 rows and 4 columns. The number of columns and rows will be adjusted when we copy and paste the data from Excel into the data input form (Figure 13.4).

Copy and paste the data from the "data for PRESENCE" spreadsheet into the Presence/Absence Data form:

- In Excel, select all the camera data in columns A to R, **including** the first column of site names, but **not** the column headings. Press Ctrl + C to copy to the clipboard.
- Back in PRESENCE, paste in the site names as well as the data: use Edit > Paste > Paste w/sitenames (you can't use Ctrl + V to paste in PRESENCE).

Click on the Site Covars tab which has two columns.

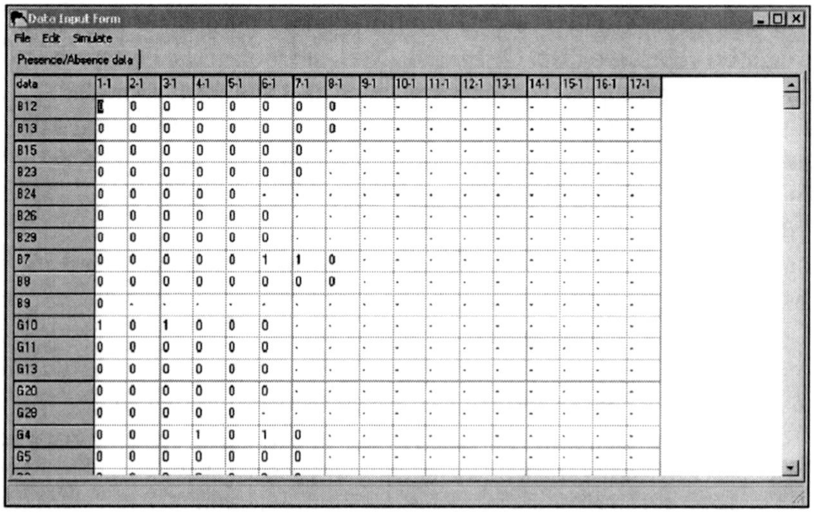

Fig. 13.4

Fig. 13.5

- In Excel, select columns T and U, **including** the column headings, and copy to the clipboard (Figure 13.5).
- In PRESENCE, use Edit > Paste > Paste with covnames.

⚠ **Save** the data to a file before you close the Input Data Form, or you may have to start over! Select File > Save as from the pull-down menus and save it as a .pao file (eg. Golden cats.pao).

Now close the form (select File > Close or click on the ⊠ button in the top right corner.)

Back at the Specification window, use the Click to Select File button and browse to the .pao file you just saved. The name of the output file (the same name with a .pa2 extension), should then appear. The full file name and path appear in the "Title for this set of data": change that to something more succinct, such as "Golden cats".

You will see that the values for No. Sites = 162, No. Occasions = 17, No. Site Covariates = 2, and No. Sampling Covariates = 0 have been updated to match the data we have just entered. The box "No. Occasions/season" is unchanged, as we do not have data for more than one season.

Click on OK. A pop-up box will appear asking for a file name: just press OK to save it again with the same file name (Golden Cats.pao or whatever you chose) and then confirm that you want to over-write the old file. (The only thing new in this file is the title for this set of data that you entered.)

PRESENCE will read your data file and a Results Browser window will appear – with no results in it until we run some analyses.

⚠ Do not close the Results Browser until you have finished the session, as the main PRESENCE window will also close without warning. If you do close down by mistake, don't panic: your work will be saved automatically and you can restart PRESENCE and reload the project.

A Simple Analysis

Select Run > Analysis:single-season, from the pull-down menu to open the Setup Numerical Estimation Run window.

We'll begin with the default model, i.e. we won't change anything, but we will look at the setup first:

Click on Custom in the Models section and the Design Matrix will appear. This looks like a small spreadsheet with tabs for Occupancy and Detection. The Occupancy matrix has one row, labeled "psi", and one column, labeled "a1". The probability that a site is occupied is usually represented by ψ, the Greek letter "psi". a1 is a parameter that PRESENCE will calculate. This model assumes that all the sites have the same probability of occupancy, so PRESENCE only needs to calculate one number and then $psi \sim 1 \times a1$. The Detection matrix also has one column, labeled "b1", which is again a parameter which PRESENCE will calculate. It has rows for each of the survey occasions, and the column is filled with 1's. This model sets all the detection probabilities for all the surveys to the same value, $p \sim 1 \times b1$.

Note that we use a $\sim$ sign, meaning "varies as", instead of $=$. PRESENCE is going to juggle the variables a1 and b2 to get the best fit to the data, but *p* and *psi* are probabilities, and can only vary between 0 and 1. To make sure *p* and *psi* stay in the right range, a **logit link** is used:

$$\log_e (psi/(1- psi)) = 1 \times a1$$
$$\log_e (p/(1-p)) = 1 \times b1$$

where *psi* $= 0$ corresponds to $1 \times a1 = -\infty$ and *psi* $= 1$ corresponds to $1 \times a1 = +\infty$, and the same for *p*.

PRESENCE works with the terms on the right hand side of the equation (called the **linear predictor**) and then translates the results back into values for *psi* and *p* in the model output.

Return to the Setup Numerical Estimation Run window, and you'll see that the name "psi(.) p(.)" has been entered for the default model. We'll see what that means later. Make sure there's something in the Title for Analysis box: type in "Golden Cats" if necessary. Leave all the tick-boxes blank and click OK to Run.

PRESENCE now juggles around with a1 and b1, finding the likelihood of getting this particular set of presence/absence data for each combination, and selecting the combination which gives the maximum likelihood. When a small box appears asking if you want to add the results to the Results Browser, click Yes.

The psi(.) p(.) model results appear in the Browser. On the right is the likelihood of the values of a1 and b1 selected based on this set of presence/absence data, expressed as $-2 \times \log(\text{Likelihood})$; it's 403.73.

(If you're not familiar with the concepts of "likelihood" and "maximum likelihood estimation", take a look at Box 13.1 towards the end of this chapter.)

Next to the log(Likelihood) is the Number of Parameters used: it's 2 (a1 and b1 in the Design Matrix). On the left is the AIC or Akaike Information Criterion, which is

$$-2 \times \log (\text{Likelihood}) + 2 \times \text{No. of Parameters}$$

The AIC and the other numbers in the table are useful when we want to compare models, so let's run a few more before we explain what they are. (And you thought PRESENCE was going to tell you how many of the camera trap sites were occupied by golden cats? Well it does, but almost as an after-thought!)

Click on the model name in the Results Browser to highlight it, then right-click and select View Model Output. A Notepad window opens with all the gory details. The first part summarizes what you put in, including the Design Matrices. Scroll down to the section headed Custom Model, where you'll find:

$$\text{Naïve estimate} = 0.2160$$

Golden cats were photographed at 35 of the 162 camera trap locations, so the **naïve estimate** of occupancy is $35/162 = 0.2160$. But that assumes that they were absent from the other 127 locations, when there's a distinct chance that one or more may have golden cats which went undetected. Towards the bottom of the file, you'll find:

```
Individual Site estimates of Psi:
   Site      Survey        Psi        Std.err      95% conf. interval
1   B12    1     1-1:    0.347150    0.065712     0.218355-0.475945
=======================================================================

Individual Site estimates of p:
   Site      Survey         p         Std.err      95% conf. interval
1   B12    1     1-1:    0.139411    0.027613     0.085290-0.193531
```

Don't be fooled by the "Individual site estimates..."! We told PRESENCE to calculate one value of *psi* and one value of *p* for all sites, and that's what it's done. These are the results for site B12 – which happens to be first on the list – and the other sites are all the same.

PRESENCE's best estimate of occupancy (*psi*) is 0.347, with 95% confidence interval (95% CI) of 0.218 to 0.476. This is higher than the naïve estimate, as some sites where no golden cats were detected were probably occupied.

The probability of detection (*p*) is 0.139 with 95% CI of 0.085 to 0.193; this refers to the probability of photographing a golden cat at least once in 5 days (if the site is occupied), and that's quite low.

With just 2 parameters, this is a bit too simple, but it gets more interesting with more complex models, which we'll now run.

More Complex Models

Select Run > Analysis:single-season from the pull-down menu to open the Setup Numerical Estimation Run window again. If Custom is selected, the Design Matrix should also open, but it might be hidden behind the Setup... window.

In the Occupancy tab, we want to tell PRESENCE to calculate separate occupancy parameters and separate detection probability parameters, one of each for "primary" and "logged" habitats (Figure 13.6A).

Make an extra column with Edit > Add col. (If the Edit menu is not visible at the top of the Design Matrix window, right-click anywhere in the window to bring it up.)

A

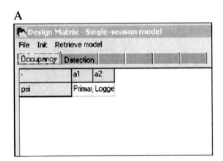

B

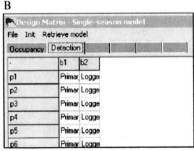

Fig. 13.6

Click in the first cell in the matrix (it already has a 1), and select Init > * Primary from the pull-down menus. In the same way select Init >* Logged for the second cell. Do the same in the Detection tab; note that Init will fill the whole column each time (Figure 13.6B).

We're now asking PRESENCE to model *psi* using two parameters, a1and a2, where

$$psi \sim a1 \times Primary + a2 \times Logged$$

and also to model *p* using two parameters, b1 and b2, where

$$p \sim b1 \times Primary + b2 \times Logged$$

If you look back at the columns in the Excel spreadsheet, you'll see that Primary = 1 for the primary forest sites, and 0 otherwise, so a1 and b1 apply only to the primary forest sites. Similarly, a2 and b2 apply only to the logged forest sites. PRESENCE will thus calculate separate *psi* and *p* values for each of the two habitat types, just as if we had pulled out the data for each habitat and run a separate analysis for each.

Go back to the Setup ... window and change the model name (which is still "psi(.) p(.)") to "psi(habitat) p(habitat)". This name indicates that we have put in habitat covariates for both psi and p. In the first model, which we named "psi(.) p(.)," there were no covariates.

Click on OK to Run and let's see what we get. Once the model has run, which may take a few moments, it will appear in the Results Browser below psi(.) p(.). PRESENCE automatically puts the model with the lowest AIC at the top of the list. The best model in the list has the lowest AIC, i.e., the lowest value for $-2 \times$ log(Likelihood) $+ 2 \times$ No. of Parameters. If you aren't familiar with AIC take a look at Box 13.1 towards the end of this exercise.

Left-click in the psi(habitat) p(habitat) row to highlight it, then right-click and select View Model Output to see the detailed output.

⚠ Note that the results displayed are those for the model highlighted, not necessarily the one you right-clicked! Check the model description at the top of the Notepad file.

The naïve estimate is of course the same, 0.2160.

```
Individual Site estimates of Psi:

    Site        Survey        Psi        Std.err      95% conf. interval
 1  B12      1   1-1:      0.299109     0.096694     0.109589-0.488629
70  A10      1   1-1:      0.380098     0.088565     0.206510-0.553686
```

Again, don't be put off by the Individual site... heading: The value for B12 applies to all the primary forest sites and A10 to all the logged forest sites.

Let's look at the values for detection probability:

```
Individual Site estimates of p:

    Site        Survey         p         Std.err      95% conf. interval
 1  B12      1   1-1:      0.146313     0.052751     0.042921-0.249705
70  A10      1   1-1:      0.136687     0.032326     0.073328-0.200047
```

Let's try the remaining two possibilities, a single estimate of psi for both habitats but separate *p*s, and a single estimate of *p* with separate *psi*s.

Select Run > Analysis:Single-Season again; Custom should already be selected and the Design Matrix should appear (if it doesn't, click on Pre-define then on Custom again.)

Select Retrieve Model > Refresh model-list from results, from the pull-down menu at the top of the Design Matrix window, then again Retrieve Model > %psi(habitat) p(habitat): the setting for that model will appear in the design matrix.

Go to the Occupancy tab and select Edit > Del Col, so there is only one column left. Then use Init > Constant'and the column will fill with 1s. Now return to the Setup... window, name the model "psi(.) p(habitat)," and run it.

Repeat the above steps for the Detection tab, leaving just a column of 1s. This is the psi(habitat) p(.) model.

We have four models in the Results Browser, which should now look like the screen shot shown in Figure 13.7.

Check the detailed output from the two new models – there don't seem to be any problems with the analysis. I've summarized *psi* and *p* (and their SEs) for all four models in Table 13.1 below. Looking at the screen shot in Figure 13.7 above, the psi(.) p(.) model is the winner, but the other models are not far behind. The AIC-weights can be used to calculate weighted model average parameters, as shown in the last line of the table above.

Assessing Model Fit

AIC and its relatives in the Results Browser only tell us which of the models we have run is the best. But it might be the best of a bad bunch: it is possible that none of the models is anywhere near a good fit. We need to check the fit of the model which

Results Browser

Model	AIC	deltaAIC	AIC wgt	Model Likelihood	no.Par.	-2*LogLike
psi(.)p(.)	407.73	0.00	0.4975	1.0000	2	403.73
psi(habitat)p(.)	409.31	1.58	0.2258	0.4538	3	403.31
psi(.)p(habitat)	409.63	1.90	0.1924	0.3867	3	403.63
psi(habitat)p(habitat)	411.28	3.55	0.0843	0.1695	4	403.28

Fig. 13.7

Table 13.1 Estimates of *psi* and *p* from the PRESENCE output (with standard errors).

Model	Psi		p		deltaAIC	AICwgt
	primary	logged	primary	logged		
psi(.) p(.)	0.347		0.139		0	0.4975
	(0.066)		(0.028)			
psi(habitat) p(.)	0.308	0.376	0.139		1.58	0.2258
	(0.084)	(0.082)	(0.028)			
psi(.) p(habitat)	0.351		0.128	0.143	1.90	0.1924
	(0.068)		(0.042)	(0.031)		
psi(habitat) p(habitat)	0.299	0.380	0.146	0.137	3.55	0.0843
	(0.097)	(0.89)	(0.053)	(0.032)		
weighted mean	0.335	0.357	0.137	0.140		

contains all the covariates – in our case, psi(habitat) p(habitat); all the models with fewer covariates are 'nested' within this one. If the top-level model fits the data well, the others will also be okay.

PRESENCE has an option to Assess Model Fit in the Setup... window. Run the model again, but this time check the box next to Assess Model Fit. You will need to give it a different name, eg. add GOF to the end; PRESENCE won't let you overwrite the old model results.

If you have a fast machine, increase the number of bootstrap replications to 500; on my laptop that took 100 seconds to run. (If you need to interrupt PRESENCE, an effective but brutal way to do it is to select the ▇ icon on the Windows task bar and press Ctrl+C.)

Add the GOF model to the Results Browser and delete the old *psi*(habitat) *p*(habitat) model: left-click to highlight it, then right-click and select Delete model. If you don't do this, PRESENCE will calculate the AIC-weights for all 5 models, without realizing that the last 2 are the same.

Look at the results for the new model in Notepad. Towards the bottom of the file you will see the calculation of the chi-squared statistic (4223.8445). This is a measure of how far the observed data are from the values calculated from our *psi*(habitat) *p*(habitat) model – it's a "badness of fit" measure; we want this to be small, but it's not obvious what's "small" and what's "big".

PRESENCE has performed a randomization test to see how many times a result this high or higher occurs if the golden cat sightings are randomly distributed. About 40–45% of the time, random values are further from the model's estimates than the actual data. (Since this is a randomization process, you will get different results.) This is not brilliant, but does mean that our model is somewhere near the actual values.

If you look at the details of the Assessing Model Fit section of the results, you will see one value in the right-hand column which is much bigger than all the others: 3663.92. Since the test statistic is simply the sum of the right-hand column, 87% of the "badness of fit" is due to this single site. If you compare the capture history (1010000101-------) with the data in the Excel spreadsheet, you'll see that this is site L3. What's going on?

- At site L3 we detected golden cats on 4 occasions out of 10. If the detection probability is the same as other sites, around 0.14, the probability of this happening by chance[2] is 0.033, about 1 in 30. Such a result is not very improbable when we have 162 cameras out there; this *might* have happened by chance. On the other hand, we might have inadvertently placed this camera near a particular cat's den; it would have been better if we had moved the camera between each 5-day period, even if only by a couple of hundred meters, so that the observations are more spatially independent.
- The chi-square value for site L3 is huge, not because the observed value is very different from the expected value (1 vs. 0.0003), but because the difference is then divided by the expected value, 0.0003. The chi-squared test is known to give biased results[3] if the expected value is low; opinions vary on what is 'low' – from 10 down to 1 – but 0.0003 is low by any standard!

If you have time, run the analysis again without site L3. You don't need to cut and paste all the data from Excel again: you can directly edit PRESENCE's input file:

---

[2] In Excel, use = BINOMDIST(4, 10, 0.14, 0), or dbinom(4, 10, 0.14) in R.
[3] See, for example, Zar (1999).

Close PRESENCE. Find the file Golden cats.pao in the folder used by PRESENCE. Open it in Notepad or other text editor. Save it with a different name – use File > Save As... and name it something like "Golden cats -L3.pao".

Change the total number of sites on the first line from 162 to 161. Now delete each of the rows referring to site L3:

- Scroll down until you find the line 1 0 1 0 0 0 0 1 0 1 - - - - - - - and delete it
- Further down you will find the heading Primary; below that delete any row with 0
- Down again and you'll see Logged; go on down and delete any row with 1
- Delete the row with L3.

Finally, right at the bottom of the file, delete the row with the number 62792; that's a check sum with the total number of entries, which is no longer valid.

Now open PRESENCE again, select File > New Project... In the Enter Specifications... window, click on Click to select file and select Golden cats -L3.pao. The modified data file should load immediately. Click OK. We are only interested in the goodness of fit test for the psi(habitat)p(habitat) model, so set that up as on page 114, check the box next to Assess Model Fit and run the model. When it has run, look at the results of the test and compare with the previous values.

Getting Finished

PRESENCE automatically saves all the results of analyses when you run them in a .pa2 file. You do not need to save results manually. To exit PRESENCE, select File > Exit or press Alt + F4 or click on the ☒ button at the top right of the window.

You can re-open the project by double-clicking on the .pa2 file name in File Manager or My Documents. You can also use File > Open project in PRESENCE, but beware!

⚠ Opening a second project with File > Open project when one is already open results in a total muddle.

To open a second project when one is already open, double-click on the .pa2 file name and it will open in a separate PRESENCE *main window*. You have to remember which is which, because there's nothing in the PRESENCE window which tells you which project file is open!

Simulating Data for Study Design

PRESENCE allows you to investigate the likely results of different study designs and to compare them with the 'true' values underlying the simulated data. It's possible to calculate how many sites and sampling occasions you must have to achieve the precision you need: see MacKenzie and Royle (2005) and chapter 6 in MacKenzie et al.'s (2006) book. However, it's still worthwhile running simulations, as it reveals some of the risks.

Open PRESENCE and select Run > Simulations from the pull-down menus. Just to see how it works, run a simulation with the default values: click on Simulate. After a few moments, a Notepad window opens with the simulation results. The first section of the file recaps the elements of the study design:

```
Total number of sites sampled:200
Number of sites sampled more intensively:100
Number of visits to intensively sampled sites:9
Number of visits to other sites:5
```

This is a pretty intensive study, involving $900 + 500 = 1400$ surveys. Next is a report on the simulation carried out:

```
Number of simulations:100
Number of times species was not detected at any site:0
Number of times convergence not achieved:0
```

You would clearly be in trouble if your study failed to detect any animals at all, or if there was a high chance of not getting enough data for PRESENCE to be able to calculate the parameters (which is what failure to converge means in practice).

Then come the real detection probability (0.2) and occupancy rate (0.7) used to generate the simulated data, and finally the results of the analysis. Since the process is based on random numbers, your results will be somewhat different from the following:

```
Average naive estimate from single visit:0.143300
Average estimate of occupancy probability:0.706045
Simulation based estimate of standard error:0.050459
Average estimate of the standard error:0.052302
```

In this case the average estimate of occupancy is very close to the true value, but that will almost always be the case if you do a high enough number of simulations. Since your study will only have one "run", you need an indication of how far from the true value you could be. The **simulation-based estimate** of standard error indicates the range of occupancy probabilities obtained during the simulation – approx. 95% of the results fall within $2 \times SE$ of the average, in this case 0.7 ± 0.1. The "average estimate of the SE" is an indication of the SE you are likely to get when you analyze the data from your single run.

Now let's run a simulation of our own. In particular, let's use simulations to investigate golden cats in plantations.

- To begin with we need plausible estimates of the true occupancy and probability of detection, and that may be a major difficulty unless a similar study has been done before or you have other data sets which you can work on, e.g., going through camera trapping records or line transect surveys to get an idea of detection probability. For golden cats in plantations, let's just assume that the true occupancy and detection rates are about the same as those for forests, Psi = 0.33 and p = 0.14.
- The camera-trapping study included 8 sites in plantations, and cameras were out for about 30 days = 6 occasions of 5 days each. Put these values into PRESENCE (see Figure 13.8).The exact results will vary between runs, as they are based on random numbers. The first thing I noticed when I looked at the results was:

```
Number of times species was not detected at any site:70
```

Fully 70% of simulations had no detections in plantations, even though the true occupancy rate was the same as that we found for forests! With this design, the study lacks the power to estimate occupancy in plantations.The average estimate of occupancy, 0.60, is almost double the true value, so even if you were lucky enough to detect golden cats, your estimate of occupancy would likely be far too high.

- Increase the number of sites to 80 (ie. roughly the same as for the Primary and Logged habitats) and try again.

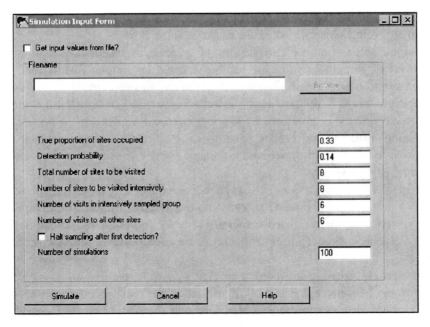

Fig. 13.8

Expected Products

- A presentation (in written, verbal, or presentation form as your instructor prefers) of your conclusions about golden cat occupancy in primary forest, logged forest and plantations. PRESENCE has produced several different estimates, and you should consider the biological plausibility of the different models as well as AIC values and other numerical output
- Responses in a form indicated by your instructor to the Discussion questions below.

Discussion

1 This survey was designed to photograph tigers for a mark-recapture study of tigers, as described in exercise 14, so it is not ideal for monitoring occupancy. What changes would you make to the design if the main objective were to monitor golden cats?
2 Think of a few species in your own locality which you may want to monitor. For which species would monitoring occupancy be appropriate? For which species would it be better to monitor overall population using line transects in chapter 12? Are there some species for which neither approach would be suitable?

Making It Happen

PRESENCE software was originally developed to analyse data on anuran calls (Spring peepers and American toads) in wetlands (MacKenzie et al., 2002) and you could implement a similar study, using frogs in wetlands or any other animal with a loud and easily identifiable call - cicadas, crickets, some birds.

The technique lends itself well to studies of animals in fragmented habitats, and the survival of animals in remnant patches created by human disturbance is important information for assessing impacts. Most urban and residential areas – and college campuses, too – have patches of woodland, wetland or grassland which support relatively isolated populations of various species. Populations do not need to be completely isolated: movement of individual animals between patches is acceptable provided new patches are not colonised by the species or old patches abandoned during your survey.

It's important that occupancy doesn't change during your study; it helps if you can complete it in a short period of time. One way to do that is to use multiple observers: if you put each observer's presence/absence data in a separate column in the spreadsheet, PRESENCE can check the detection probabilities for each observer and correct for differences.

If the same sites are visited in subsequent years, it's possible to check on changes in overall occupancy. Moreover, PRESENCE can calculate rates of colonization of patches and rates of local extinction: see MacKenzie et al. (2003).

The techniques can be used too on a small scale to estimate the proportion of rose bushes occupied by aphids or cabbages occupied by caterpillars.

Further Resources

The Help facility in PRESENCE is a bit sketchy, with only an Overview. However, the Help menu also includes good tutorials using data sets which come with the package and are the same analyses discussed in MacKenzie et al. (2006).

The main reference is the book by MacKenzie et al. (2006), which nicely links ecological examples to the theory. Darryl MacKenzie with various co-authors has produced a range of papers on individual aspects of occupancy and the use of PRESENCE.

Box 13.1   Likelihood, Maximum Likelihood Estimation and AIC

Recall that in the section entitled "Occupancy of discrete habitat patches" we had 10 ponds which we knew had frogs, but we only detected frogs in 6 of them. We concluded that "the probability of detecting frogs when they are there is 0.6." But is that the right value? And if so, why is it the right value?

To understand those questions better, think of tossing a coin 10 times and getting 6 heads. Would you conclude that the probability of getting a head, $p(\text{head}) = 0.6$? We know enough about coins to be sure that $p(\text{head}) = 0.5$, and we also know that getting 6 heads out of 10 is not unusual. The probability of getting various numbers of heads out of 10 is shown in the table and graph below (Table 13.2 and Figure 13.9). (You can calculate these using dbinom(6, 10, 0.5) in R or BINOMDIST(6, 10, 0.5, 0) in Excel.)

In the same way, we would be quite likely to detect frogs at 6 out of 10 ponds if the probability of detection, $p(\text{detect}) = 0.5$. We can't be certain that

(Continued)

Box 13.1 (***Continued***)

121

Analyzing Camera Trap Data

Table 13.2

no. of heads (x)	Pr (x heads/10) given p(heads = 0.5)
0	0.00098
1	0.00977
2	0.04394
3	0.11719
4	0.2050
5	0.24609
6	0.20508
7	0.11719
8	0.04394
9	0.00977
10	0.00098

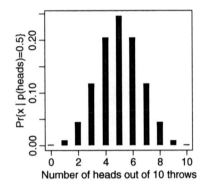

Fig. 13.9

p(detect) = 0.6, but that might be the best estimate we can make, based on the data we have.

Now let's calculate the probability of getting 6 detections out of 10 with different values of p(detect), as shown in Table 13.3.

Notice that we use a different term in the table and graph, **likelihood** (and the symbol $\mathscr{L}$), although we still use probability functions to calculate its value. For given values of p(detect) and number of detections, likelihood and probability are the same;[4] for example:

$$\mathscr{L}\{p(\text{detect}) = 0.6|6 \text{ detections out of } 10\} = \Pr\{6 \text{ detections out of } 10 \,|\, p(\text{detect}) = 0.6\}$$
$$= 0.2508$$

But probability and likelihood have different interpretations, as you will see from the graphs above. We use the term **probability** when we are interested in a range

Table 13.3

p(detect)	$\mathscr{L}$ (p given 6 detections/10)
0.00	0
0.10	0.0001
0.50	0.2050
0.59	0.2503
0.60	**0.2508**
0.61	0.2502
0.70	0.2001
0.90	0.0111
0.95	0.0009
1	0

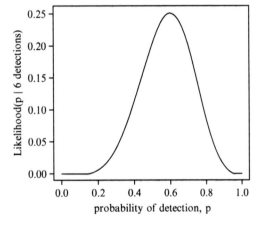

Fig. 13.10

[4] Strictly speaking, they are proportional, ie. $\mathscr{L} = Pr \times$ constant, but all statisticians agree that the sensible value for the constant is 1.

(*Continued*)

of *observations*, e.g. number of detections = 0 or 1 or 2 or 3 ... for a fixed *p*(detect). We use **likelihood** for a range of values of a *parameter*, e.g. *p*(detect), for a fixed observation (e.g. number of detections = 6).

The best **estimate** of *p*(detect) is the value with the **maximum likelihood** given the particular data set that we collected. As you can see from the table and graph, this is *p*(detect) = 0.60. This is the **Maximum Likelihood Estimate** and is represented by the symbol $\hat{p}$ (p-hat): $\hat{p}$(detect) = 0.60.

The answers to the two questions in the first paragraph are:

- 0.6 may not be the real value of the probability of detection, but it is the best estimate we can make based on the data we have.
- It is the best estimate because the probability of detecting frogs in exactly 6 ponds out of 10 is higher for *p*(detect) = 0.6 than for any other value.

Now suppose we collect more data. We visit the 10 ponds with frogs again, and this time we detect frogs at only 5 ponds, even though we have no reason to suspect that *p*(detect) has changed. To find the likelihood of *p*(detect) given that we detected frogs at 6 then 5 ponds, we multiply together the individual likelihoods:

$$\mathscr{L}(p|6 \text{ then } 5 \text{ detections}/10) = \mathscr{L}(p|6 \text{ detections}) \times \mathscr{L}(p|5 \text{ detections})$$

As you can see from Table 13.4, the Maximum Likelihood Estimate of *p*(detect), $\hat{p}$ (detect) = 0.55. The true value of *p*(detect) may not have changed, but our estimate, $\hat{p}$ (detect), has changed to reflect the additional data.

You will also see that the likelihood gets smaller and smaller each time we add more data. With lots of data, the likelihood is extremely small. For example, the likelihood for the psi(.) p(.) model for the golden cats analysis in PRESENCE is:

```
0.00000000000000000000000000000000000000000000000000000000000000
0000000000000000002185
```

So we prefer to work with the (natural) log of the likelihood, which in this case is −201.86. In fact, the result displayed in the Results Browser in PRESENCE is "−2 ∗ LogLike", which is 403.73.

Using logs makes no difference to Maximum Likelihood Estimation, as the maximum of the log likelihood always corresponds to the maximum likelihood.

To summarize so far: If *p*(detect) hasn't changed, $\hat{p}$(detect) = 0.55, the likelihood of this value is 0.0558 (last column in the table above) and the log likelihood is $\log_e$ (0.0558) = −2.886.

Table 13.4

p(detect)	$\mathscr{L}(p \mid 6 \text{ detections}/10)$	$\mathscr{L}(p \mid 5 \text{ detections}/10)$	$\mathscr{L}(p \mid 6 \text{ then } 5 / 10)$
0.49	0.1966	0.2456	0.0483
0.50	0.2051	**0.2461**	0.0505
0.51	0.2130	0.2456	0.0523
0.54	0.2331	0.2383	0.0556
0.55	0.2384	0.2340	**0.0558**
0.56	0.2427	0.2289	0.0556
0.59	0.2503	0.2087	0.0522
0.60	**0.2508**	0.2007	0.0503
0.61	0.2503	0.1920	0.0481

(*Continued*)

Box 13.1 (*Continued*)

123

Analyzing Camera Trap Data

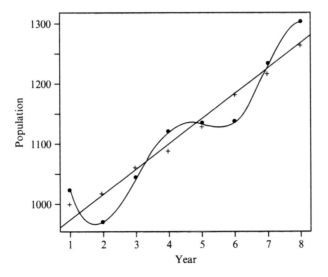

Fig. 13.11

Now let's consider a second model. If p(detect) has changed, we can estimate different values for each occasion, say p_1 and p_2. On the first occasion we detected frogs at 6 out of 10 ponds, so $\hat{p}_1 = 0.60$ with likelihood 0.2508 (column 2 in Table 13.4). On the second occasion we detected frogs at 5 out of 10 ponds, so $\hat{p}_2 = 0.50$ with likelihood 0.2461 (column 3 in the table). The likelihood for this model is $0.2508 \times 0.2461 = 0.0617$ and log likelihood $= -2.785$.

Comparing the two models, we see that the likelihood for model 2 is higher than for model 1 (0.0617 vs 0.0558), which means that it is a closer fit to the data we collected. But model 2 is also more complex, having two parameters (p_1 and p_2) instead of just one. The model which is the best fit to the *data* is not necessarily the best representation of *reality*.

To see the difference, let's look at a different situation. The graph shows the numbers for a hypothetical population of elephants for eight years. The crosses indicate the true population, which has an underlying upward trend. The black dots are the estimates from annual surveys of elephants. The lines represent two models calculated from the survey data. The straight line is a simple model with just 2 parameters; the wiggly line has 8 parameters and fits the data perfectly. In fact, the "wiggliness" of the wiggly line is due mainly to errors in estimating the population, and the simpler model is closer to the true situation. The wiggly model is described as "over-fitted" or "over-parameterized."

In reality of course we don't know the true population, all we have are estimates which are never quite exact – we see the dots but not the crosses in the diagram. How do we decide which is the best model?

Akaike's Information Criterion (AIC) balances number of parameters and fit to the data (likelihood):

$$\text{AIC} = 2 \times \text{No. of Parameters} - 2 \times \log(\text{Likelihood})$$

and a small value for AIC indicates a better combination of simplicity and fit to data. As shown in Table 13.5 below, Model 1 has a smaller AIC value, and is the better model by this criterion.

(*Continued*)

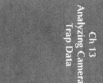

Box 13.1 (*Continued*)

Table 13.5

	Model	No. of Parameters	Likelihood	log (Likelihood)	AIC
Model 1	$p_1 = p_2 = p$	1 ($\hat{p} = 0.55$)	0.0558	−2.886	2 + 5.762 = 7.762
Model 2	$p_1 \neq p_2$	2 ($\hat{p}1 = 0.60$; $\hat{p}2 = 0.50$)	0.0617	−2.785	4 + 5.570 = 9.570

Table 13.6

	Model	AIC	deltaAIC	Model Likelihood	AIC weight
Model 1	$p_1 = p_2 = p$	7.762	0	1	1/1.40 = 0.71
Model 2	$p_1 \neq p_2$	9.570	9.570 − 7.762 = 1.808	0.40	0.40/1.40 = 0.29

The AIC values enable us to answer more questions about the models. Model 1 is better then Model 2, but is it a lot better, or only a little better? The difference in AIC (called **deltaAIC**) is only 1.8, which means that Model 1 is only a little better: Burnham and Anderson (2002, page 70) suggest that models with deltaAIC values less than 2 are all plausible, 4–7 are considerably less so, and more than 10 means that the models miss some important explanatory variables.

We can also ask, "How likely is it that Model 2 is in fact the correct model?" We can only calculate the *relative* likelihoods of the models, but if we take the best model as having Model Likelihood = 1, then

$$\text{Model Likelihood} = e^{-\text{deltaAIC}/2}$$

For Model 2, that works out to 0.40 (see Table 13.6 above), so Model 1 is more than twice as likely as Model 2 to be the right one, but Model 2 can't be ruled out.

If we were quite sure that Model 1 was right, we would be confident in our estimates of detection probabilities, $\hat{p}_1 = \hat{p}_2 = \hat{p} = 0.55$. But if we can't rule out Model 2, the right values might be $\hat{p}_1 = 0.60$ and $\hat{p}_2 = 0.50$. We really need an estimate which takes into account the uncertainly about the best model.

The solution is **model averaging**: we take a weighted average of the estimates from the different models, using weights proportional to the Model Likelihood. As shown in the last column of the table above, AIC weights are simply the Model Likelihood divided by the total of all the Model Likelihoods. For our example, this gives:

$$\hat{p}_1 = 0.71 \times 0.55 + 0.29 \times 0.60 = 0.5645$$
$$\hat{p}_2 = 0.71 \times 0.55 + 0.29 \times 0.50 = 0.5355$$

AIC is a powerful way of working with models and is used by DISTANCE and MARK as well as PRESENCE. It can be used to compare models which are completely different in form. But you must use the same data set for all the comparisons. For example, the results for site L3 in the golden cat data look suspicious, so we might try deleting that site and reanalyzing the data; we can't use AIC to compare the models with and without L3.

For more on modeling and model fitting with AIC, see Burnham and Anderson (2002).

CHAPTER 14

Estimating Population Size with Mark-recapture Data and MARK

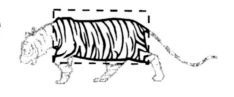

Michael E. Meredith

Most wild animals are elusive, but we still need to know basic facts about their populations. Conservation biologists are frequently asked "How many animals are there in this population?" and "Is this population decreasing, stable, or increasing?" But unless a particular species aggregates in an easily countable fashion, for example, seabirds assembling en masse into a nesting colony or elephants herding on an open plain, we have to resort to gleaning bits of information from occasional sightings of particular individuals to say much in general about their populations.

Occasional sightings (and re-sightings) of particular individuals is the basis of **mark-recapture** studies of populations. Generally speaking, a researcher visits a study area and captures a number of individual animals. Each individual is marked with a unique identifier (for example, colored bands for birds or an ear tag for a mammal). For some species we might recognize individuals from photographs or field description based on distinctive coloration patterns so that we do not need to actually trap and handle them (usually best for them and occasionally for us, as well, as in the case of large carnivores!). We then release all marked individuals back into the environment unharmed. Allowing sufficient time to pass for the marked individuals to redistribute themselves among the unmarked population, we then return and capture another sample of individuals. Some will have been marked during the initial visit (these are "recaptures"). Others will be new and are marked and released during the second visit.

With just two visits we can estimate population size from the ratios of previously marked to newly encountered, unmarked animals. Repeating this sampling process for three or more sampling occasions permits us to assemble "capture histories" for particular individuals, which lets us estimate population size with even greater precision but also lets us make estimates of survival or movement.

Analyzing capture histories relies on sophisticated software but the insights provided on population size, survival, or movement are worth the effort required to gain competency with the software that does the analysis. In fact, for many elusive species of conservation concern, mark-recapture analysis is the only option available to us to answer critical questions about population size and survival rates. This chapter provides you the opportunity to become familiar with the basic concepts behind mark-recapture analysis as well as to "test-drive" the primary software package available for doing these kinds of analysis.

Objective

To understand the concepts behind population estimation with mark-recapture data and to gain experience of using the supporting software, MARK.

Procedures

Basic Concepts

We'll start with a simple experiment to show how capture-mark-recapture methods enable us to estimate the size of a population of animals. For this experiment you'll need a deck of standard playing cards, or approximately 50 identical cards, such as business cards or index cards.

Step 1 : Deal the cards into, say, 20 small stacks on the table. "Capture" the card at the top of each stack. If your cards are blank, mark the captured cards with the letter A; this is analogous to trapping animals such as small rodents, which need an artificial mark, such as clipping short the fur on the left shoulder. If you are using playing cards, list the cards you have captured; this is analogous to capturing animals that can be individually identified from natural markings, such as tigers' stripes, leopards' spots, or elephants' ear shapes.

Step 2 : "Release" the captured cards to the stacks where they were caught. Gather up the stacks, shuffle the cards well, and deal again into small stacks. The number of animals you capture this time may not be the same as the number you captured the first time, so now use, say, 16 stacks. "Capture" the card at the top of each stack. Count the number of cards that are "recaptures".

Now we can estimate the total number of cards. We take the proportion of recaptures in the second sample as an estimate of the proportion of marked cards in the whole pack. Since we know how many cards we marked, we can calculate how many are in the pack.

When I did this, I marked 20 cards the first time; the second time, I captured 7 marked cards, so my estimate of capture probability is $7/20 = 0.35$, i.e. I've captured about 1/3 of the marked individuals. If the capture probability is the same for marked and unmarked cards, then the 16 cards in the second sample is about 1/3 of the total number of cards, so the total is about $16 \times 3 = 48$. More exactly:

$$N = 16 \times 20/7 = 45.7$$

When I counted the cards, there were exactly 50, so my estimate was off by about 10%.

This is the Petersen method, and it depends on a couple of assumptions (which obviously apply to the experiment with the cards, but might not apply if we were trapping real animals):

1 The population is closed: no additions (births or immigration) and no losses (deaths or emigration) occur between the two sampling occasions.

2 Capture probability is the same for marked and unmarked animals. In practice, animals which have been trapped once (and marked) often become 'trap-shy' and avoid traps in future. Occasionally – especially if traps are baited – animals become 'trap-happy' and more likely to be captured in future.

You could repeat step 2 of the exercise and compare the different estimates you get. But if we are going to have more than two capture sessions, there are better ways of combining the results. First, though, we need to do a "capture history" for the first two sessions.

We use 1 to indicate capture and 0 for non-capture, so '1 1' means captured twice, '1 0' means captured the first time but not the second, and '0 1' means captured the second time but not the first. My second trapping session resulted in recapturing 7 out of the 20 cards captured the first time, so:

$$1\ 1 = 7$$
$$1\ 0 = 20 - 7 = 13$$

There were also 9 new captures:

$$0\ 1 = 9$$

If you are using playing cards, you can use a different scheme, as described below.

Step 2(a) : Do the capture history for your experiment so far:

$$1\ 1 =$$
$$1\ 0 =$$
$$0\ 1 =$$

With playing cards, you can keep capture histories for individual cards, for example, if the jack of clubs was captured the first time but not the second, and the 3 of hearts was captured both times:

$$JC = 1\ 0$$
$$3H = 1\ 1$$

Step 3 : Mark all the cards you captured in step 2 with the letter 'B' or, if you're using playing cards, add the extra cards to your list with capture history '0 1'. "Release" the captured cards, gather up the stacks and shuffle, then deal again – I used 22 small stacks this time. Update the capture history.

My pack already contained 7 cards with both 'A' and 'B' (the '1 1 = 7' in the old capture history) and I caught 5 of them this time, so there must be $7 - 5 = 2$ that I didn't catch, hence:

$$1\ 1\ 1 = 5$$
$$1\ 1\ 0 = 7 - 5 = 2$$

I caught 8 cards out of 13 with A but not B, and 2 out of 9 cards with B but not A:

$$1\ 0\ 1 = 8$$
$$1\ 0\ 0 = 13 - 8 = 5$$
$$0\ 1\ 1 = 2$$
$$0\ 1\ 0 = 9 - 2 = 7$$

And finally, 7 cards with no marks:

$$0\ 0\ 1 = 7$$

How do we analyze these results? We need to estimate two parameters, population size and capture probability, which are interrelated. To do this efficiently we need computer software, such as MARK, which we will get to know in a moment. First, though, do two more captures:

Step 4 : Mark all the cards you captured in step 3 with the letter C or, if you're using playing cards, add the extra cards to your list with capture history '0 0 1'. "Release" the captured cards, gather up the stacks and shuffle, then deal again, using a small number of stacks – I used 8. Update the capture history.

Step 5 : Mark the cards you captured with the letter D or add the extra cards to your list with capture history '0 0 0 1'. Shuffle and deal again, using 20 stacks. Update the capture history.

Put the cards to one side but keep the capture history, which you will later analyze with MARK. However, for a first run through with MARK, we'll use a real data set.

Getting Started with MARK

The best place to download the latest version of MARK is www.phidot.org/software/mark/download/. As with any .exe file, it's wise to download it, run a virus check with an up-to-date virus scanner, and create a Windows Restore Point (go to Start > Programs > Accessories > System Tools > System Restore) before running the setup program.

If you use the results from MARK in any report or paper, please cite it as:

White, G.C. and K. P. Burnham. 1999. Program MARK: Survival estimation from populations of marked animals. *Bird Study* 46, Supplement: 120–138.

The Kanha Tiger Data

As part of a country-wide survey of tigers, Ullas Karanth and his team used automatic cameras to "trap" tigers in Kanha National Park in Madhya Pradesh state in central India (Kanha is a popular tourist destination; you can get more background information about it from a web search). The cameras were set on 10 occasions totaling 803 camera-trap days. They got photos of 26 individual tigers, which were identified using their distinctive patterns of stripes (Figure 14.1).

Some tigers were "captured" on photos 2 or 3 times, some only once, and the researchers were able to deduce the capture history for each tiger. The survey was completed in three months; this is short enough for us to be able to assume that the same tigers stayed in the area for the whole time, ie. we have a "closed population" of tigers.

Fig. 14.1 Striping patterns permit individual tigers to be distinguished. From p. 149, Karanth and Nichols 2002, © Centre for Wildlife Studies.

with Notepad to see the contents. (If you can't download the file, open a new file
in Notepad, type in the data from the box below, then Save the file with the name
"Kanha_tiger.inp").

```
/* Kanha tiger data from the WCS India Program */
/* 1 */ 1001000110 1;
/* 2 */ 1000000101 1;
/* 3 */ 1101000011 1;
/* 4 */ 0110000111 1;
/* 5 */ 0101000001 1;
/* 6 */ 0100000000 1;
/* 7 */ 0010011000 1;
/* 8 */ 0011000000 1;
/* 9 */ 0001000011 1;
/* a */ 0001100110 1;
/* b */ 0001001001 1;
/* c */ 0000100000 1;
/* d */ 0001000000 1;
/* e */ 1001000000 1;
/* f */ 0001000000 1;
/* g */ 0001100000 1;
/* h */ 0001001000 1;
/* i */ 0000110000 1;
/* j */ 0000100000 1;
/* k */ 0000010000 1;
/* l */ 1000101000 1;
/* m */ 0100000000 1;
/* n */ 0001000001 1;
/* o */ 0000010000 1;
/* p */ 0000000001 1;
/* q */ 0000100000 1;
```

The first line is just a comment to confirm what data are in the file. MARK ignores
anything enclosed with /* . . . */ characters, assuming it is of interest to humans but
not to machines. The codes on the left are the IDs of the individual tigers identified
from their coat pattern. MARK will ignore these too, but they will help if we need to
check or modify the data later. Then comes the capture history for each tiger: the
traps were set up for 10 periods, and the string of ten 0s and 1s indicates whether that
tiger was photographed (1) or not (0) during each of the 10 periods. The last number
is the number of animals with that capture history; in this case, each line corresponds
to a single tiger, but (for example) tigers "d" and "f" have the same capture history,
so we could replace those two lines with a single line: "/*d and f */ 0001000000 2;".
Each line of data finishes with a semi-colon (;).

Close the Notepad file.

Setting up the Project and Entering Data in MARK

Start MARK and select File > New from the pull-down menus. In the Enter
Specifications for MARK Analysis window, under Select Data Type, click on Closed
Captures. A new Model Selection... window appears.

We use a **closed captures analysis** because we can assume that the same tigers
remained in the area throughout the survey. The Closed Captures option has 12
possible sub-options, but only 3 are relevant to us. The simplest possible situation

is when capture is equally probable for all animals on all trapping occasions: this is often called the **M0 model**. We can design more complicated models, for example:

Mb : the animal's behavior changes after being trapped once, so that capture and recapture have different probabilities

Mt : capture probability is not the same on each occasion, it varies with time

Mh : heterogeneity - capture probabilities are not the same for all animals.

And it's possible to combine two or three effects, but then it gets complicated! The three options relevant for our analysis are:

- Closed captures (the "plain vanilla" option) – for M0, Mb, Mt and Mbt (no heterogeneity)
- Closed captures with heterogeneity – for Mh (no behavior or time effect)
- Full closed captures with heterogeneity – for all models, including Mtbh.

There are corresponding "Huggins" versions, which use a rather different way of calculating the total population. There are also six "misidentification" versions, developed for use when capture–recapture is based on DNA analysis, which might not identify individual animals with certainty. It's possible to switch between models during the analysis, so we'll start with the basic one and change if we need to.

Highlight Closed captures and click on OK. In the Enter Specifications... window, type a title for the set of data (eg "Kanha tigers"). Then click on Click to Select File, browse to 'Kanha_tiger.inp' and click Open. MARK enters the name of the Results file as Kanha_tiger.dbf, but you can change that if you wish (e.g. if you are doing different sets of analyses with the same input file).

Change Encounter Occasions to 10, and leave Attribute Groups as 1. Click on OK. A dialogue box pops up to inform you that the output file kanha_tiger.dbf was created. Click OK.

Running a First Analysis in MARK

When you've entered all the specifications, a PIM (Parameter Index Matrix) window opens. There are three of these, so let's open all of them before discussing the details.

From the pull-down menus, select PIM > Open Parameter Index Matrix. In the box that appears click Select All then OK. Drag the three windows so that you can see the main items in all three (see screen shot, Figure 14.2). The three windows show the parameter indices. The default uses the maximum number of parameters, so there are:

- 10 parameters for Capture Probability (p), numbered 1–10, one for each trapping occasion
- 9 parameters for Recapture Probability (c), numbered 11–19 (only 9, because you can't recapture animals on the first trapping occasion)
- 1 parameter for Population Size (N).

In this case the model with so many parameters produces errors (try it if you like!), so let's begin with the minimum number of parameters, the M0 model, with all the capture and recapture probabilities equal.

Change all the Capture Probabilities to 1; you could type "1" in each box, but a quicker way is to click on any of the boxes and select Initial > Constant from the pull-down menus. Then change all the Recapture Probabilities to 1 as well: click anywhere in the Recapture window and select Initial > Make $c=p$ from the menus.

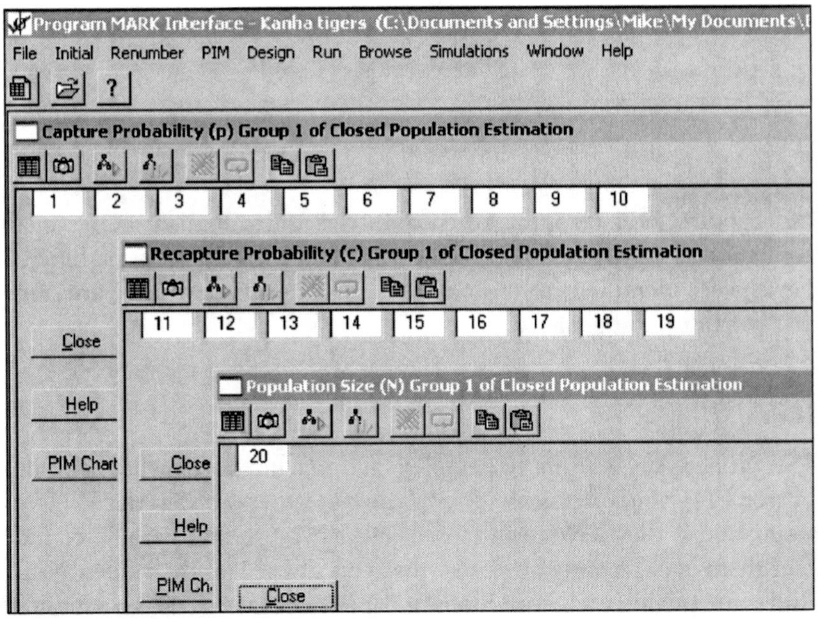

Fig. 14.2

This tells MARK to use the same Beta coefficient (β_1 in this case) for all the capture and recapture probabilities. We are not saying that the capture probability =1. It would make no difference to the result if you told MARK to use β_1 for the population size and β_2 for the capture and recapture probabilities. In the Parameter Index Matrix (PIM) we are only dealing with the indices (the little numbers identifying the βs), not with the values of the parameters.

The number in the Population Size box is still 20, but that doesn't matter, provided it is different from the other numbers.

Select Run > Current Model from the pull-down menus. In the Setup Numerical Estimation Run window, give the model the name M0, leave all the other default values as they are and click OK to Run. Click Yes when MARK asks if it should use an identity design matrix, and Yes again when it asks if it should add the output to the database.

The Results Browser appears, which will look familiar if you have used PRESENCE or DISTANCE before. There's nothing interesting in the Browser yet, as we have only run one model, but you might note that No. Parameters is 2, as we intended; these are (i) capture/recapture probability ($p = c$) and (ii) population size (N). Let's look at the detailed output.

Right-click in the Results Browser and select Output in Notepad (note that the results for the highlighted model appear, not the one you clicked on!) or click on the second button 🖽 on the toolbar. The first part summarizes the input, including the model name and setup details. There's a warning that "At least a pair of the encounter histories are duplicates"; we put each tiger on a separate line, so that is okay. Then there's some technical information and, near the bottom, the *real* stuff. Just like PRESENCE (see chapter 13), MARK deals in probabilities which vary from 0 to 1, and these cannot be modeled directly by a linear function. Like PRESENCE, MARK uses a linear estimator which is linked to the real values of p and N; in this case the default is the SIN link function (PRESENCE uses the logit link by default).

```
                    Real Function Parameters of M0
                        95% Confidence Interval
Parameter       Estimate        Standard Error  Lower           Upper
-----------     --------------  --------------  --------------  ---------
1:p             0.2038924       0.0285186       0.1536181       0.2654594
2:N             28.446382       2.1734708       26.548565       36.909891
```

The estimated capture (and recapture) probability is quite high at 0.2, and the estimated population for the part of the Reserve surveyed is 28.4 tigers. Since 26 tigers in the area were identified from the photos, that means that there are only 2 or 3 tigers that weren't photographed.

Comparing Models

Now let's see what happens with more complex (and maybe more realistic) models.

Open the three PIM windows again. If you can't see them, check the Windows pull-down menu, and if they have been closed, use PIM > Open Parameter Index Matrices to get them back. Arrange them so that you can see all the values. Notice that MARK did some tidying up before running the previous model and renumbered the Population Size parameter to 2.

Mb

Let's run a model where animals behave differently after being trapped the first time, so recapture probability is different from capture probability. But neither changes with time. In the Recapture Probability window, click on the +button: all the values will increase to 2. Do the same in the Population Size window, changing the 2 to 3. Run the analysis as before, naming the model Mb.

You could set the index for Recapture Probability to 3 and leave Population Size as 2, but the output is easier to interpret if you always have Population Size values last. In any case, do make sure that the Population size index is not the same as the Recapture index (to avoid that mistake and the nonsense results that follow, it's best to open *all* the PIM windows).

Look at the output for this model. The estimate for population size is 27, with confidence limits of 26 to 39. This is only a little higher than the results for the M0 model, and c and p are fairly similar.

Mt

Now try the model where capture and recapture probabilities are equal, but they vary with time. Open the three PIM windows again. Click in the Capture Probability window and select Initial > Time from the pull-down menu, then Initial > Make c=p.

The parameters 1 to 10 will be used to model 10 different capture probabilities for the 10 trapping occasions, and the parameters 2 to 10 will also be used for the recapture probabilities – remember that there are no *re*captures for the first trapping occasion. Change the parameter index for Population Size to 11 (or any number above 10). Run the model as before, naming it Mt.

The population estimate and its confidence limits look reasonable. The output contains 10 separate estimates for p, but many of them are the same (0.1425971 and 0.1782464 each occur three times), which suggests that MARK does not have enough data to do a good job of estimating separate values.

So far we've assumed all tigers have the same probability of being captured. MARK can't handle separate capture probabilities for each tiger, but it can work with 2 or sometimes 3 groups with different probabilities and mix the results. We'll try mixing just 2,[1] but to do that we have to change the data type.

Select PIM > Change Data Type from the pull-down menus. In the list of options that appears, highlight Closed Captures with Heterogeneity and click on OK. Leave the Number of Mixtures as 2 in the next box and click OK. This data type allows for a basic heterogeneity model equivalent to mixing two M0 models, so there is only one PIM window for p, with two indices. The parameter pi (π) determines how much of each model goes into the mix. If we can divide the population into two groups of animals with different p values, pi indicates the proportion of the population in each group. There isn't much we can change in this model, so we'll use the defaults.

Run the model as before, naming it Mh. The estimate for pi is 0.49, so nearly half of the tigers fall into each group, one with capture probability $p = 0.26$ and the other with $p = 0.11$. The estimated population – and in particular the upper confidence limit – is much higher than for the other models, which indicated populations only a little higher than the 26 animals actually captured. The Mh model tells us that some tigers have a much lower capture probability than average, so there are likely to be more tigers that we failed to photograph.

Model	AICc	Delta AICc	AICc Weight	Model Likelihood	No. Par.	Deviance
{M0}	155.2900	0.0000	0.53946	1.0000	2	119.2551
{Mb}	156.5433	1.2533	0.28828	0.5344	3	118.4614
{Mh}	158.7985	3.5085	0.09335	0.1730	4	118.6534
{Mt}	159.1343	3.8443	0.07892	0.1463	11	104.0817

Now let's look at the summary of the models in the Results Browser. MARK uses AICc: the usual Akaike Information Criterion (AIC) is negatively biased (i.e., too small) when the sample size is small compared to the number of parameters, and AICc includes a small-sample correction term.

As with DISTANCE and PRESENCE, the models are ranked according to AICc, with the lowest AICc indicating the "best" model in the sense that it is a good compromise between complexity (= number of parameters) and fit (= likelihood of getting this set of data if this model is correct). In this case, the M0 and Mb models are very close, and the other two models are not far behind. The time-dependant model, Mt, gives the best fit (lowest Deviance), but that is offset by the large number of parameters.

Combining Models

The population estimates for the four models are shown in the top four rows of Table 14.1. No model is clearly the best. In this situation we can use the AICc weights to calculate a weighted average of the four population estimates. MARK will do the sums for us:

---

[1] This was tried with 3 groups for this data set and it failed to converge.

Table 14.1 Population estimates for four MARK models.

Model	Estimated population	Confidence interval		Capture probability	delta AICc	AICc weight
		lower	upper			
MARK: M0	28.45	26.55	36.91	0.20	0	0.54
MARK: Mb	26.94	26.08	36.86	$p = 0.25$	1.25	0.29
				$c = 0.19$		
MARK: Mh	31.52	26.66	72.24	0.26, 0.11	3.51	0.09
MARK: Mt	28.05	26.41	36.18	(10 values)	3.84	0.08
Weighted average	*28.27*	*26*	*34.98*			
CAPTURE: M0	28	26	34	0.20		
CAPTURE: Mh	33	29	49	0.18		

In MARK, go to Model Output > Model Averaging > Derived. In the Model Averaging Parameter Selection window which opens, check the box next to 1, then *uncheck* the box next to Only select models for the current data type (remember that we changed the data type before calculating the Mh model). Press OK. A warning will pop up saying that averaging is not properly supported across data types. Press OK. The results appear in a Notepad window.

```
                              Derived Parameter 1
Model              Weight            Estimate              Standard Error
---------------------------------------------------------------------------
{M0}               0.53946           28.4463816            2.1734708
{Mb}               0.28828           26.9389230            1.8205530
{Mh}               0.09335           31.5205171            8.2640357
{Mt}               0.07892           28.0510578            1.9997066
---------------------------------------------------------------------------
Weighted Average                     28.2675790            2.6265523
Unconditional SE                                           3.4222897
95% CI for Weighted Average Estimate is 21.5598911 to 34.9752669
Percent of Variation Attributable to Model Variation is 41.10%
```

If you check output in Notepad for any of the models, you'll see that N occurs twice, once as a parameter which has been estimated and again as a Derived Parameter, the Population Estimate (N-hat): they are identical.

The weighted average of the estimates, 28.27, is straightforward. MARK also gives Standard Errors (SEs), calculating a Weighted Average SE and a (larger) Unconditional SE. The SEs of the individual models are valid *if* the model is correct: for example, the SE is 2.17 on condition that model M0 is the right one. The Unconditional SE reflects the SEs of the individual models, plus the additional uncertainty about which model is correct. MARK uses the Unconditional SE to calculate 95% Confidence Intervals, and has come up with an implausible value for the lower limit: we know there are at least 26 tigers in the area. The problem is that calculating a 95% CI from the SE assumes that the uncertainty is symmetrical about the point estimate. In fact, the lower limit is closer to the estimate and the upper limit is further away.

Comparison with CAPTURE

CAPTURE is an older program (from 1995) which can be run from within the MARK interface or as a stand-alone program. It analyzes closed capture data, but unlike MARK it does not rely on maximum likelihood estimation and AIC. CAPTURE offers several models and other features not available in MARK, in particular Mh models which allow a different capture probability to be estimated for each animal.

For this situation, three features of CAPTURE are important: goodness of fit testing, method of selection of models, and methods of dealing with heterogeneity.

Goodness of fit tests

Select Tests > Program CAPTURE from the pull-down menus. In the Program CAPTURE Models box, check "Appropriate" to ask CAPTURE to look for the most appropriate model. Click on OK.

A Notepad window opens with the output. It starts with a summary of the input data, followed by the 7 tests which CAPTURE carries out. Look through them, remembering that a large probability value means that we cannot reject the null hypothesis, which is that the data fit the model. Tests 1 to 3 suggest that M0 compares well with the other models (probabilities more than 0.1); Tests 4, 5 and 7 give little support for any of the other models (probabilities around 0.05); Test 6 could not be run with the present data set.

Model selection algorithm

Instead of AIC, CAPTURE combines the results of the goodness-of-fit tests to choose the best model. An algorithm developed from simulated data sets is used to calculate relative appropriateness of the different models (See Otis et al. (1978), especially p 57 ff). The appropriateness of each model is estimated as a weighted linear combination of the test results and scaled so that the most appropriate has a score of 1. These are summarized near the end of the output:

```
Model selection criteria. Model selected has maximum value.

Model      M(o)   M(h)   M(b)   M(bh)  M(t)   M(th)  M(tb)  M(tbh)
Criteria   1.00   0.87   0.22   0.52   0.00   0.27   0.40   0.64

Appropriate model probably is M(o)
Suggested estimator is null.
```

The result using the recommended model is then given:

```
Population estimation with constant probability of capture.

Estimated probability of capture, p-hat = 0.2039
Population estimate is 28 with standard error 2.1386
Approximate 95 percent confidence interval 27 to 36
```

So CAPTURE has also selected the M0 model as the most appropriate, and gives us almost the same results as the M0 model in MARK (upper confidence limit of 36 instead of 37).

Individual heterogeneity

According to CAPTURE's analysis, the Mh model is second-best and not far behind M0. The Mh model using the jackknife estimator is robust to violations of the model assumptions, whereas the M0 model tends to produce negatively biased results in the presence of heterogeneity (Otis et al, 1978). The trade-off is that Mh normally has wider confidence intervals.

CAPTURE provides several alternative Mh models, and the results for the Jackknife Mh model are:

Select Test > Program CAPTURE from the pull-down menus again, but this time check Jackknife – M(h). Click on OK.

The important bit in the Notepad file that opens is:

```
Population estimation with variable probability of capture by animal.
...
Average p-hat = 0.1758
Interpolated population estimate is 33 with standard error 4.6859
Approximate 95 percent confidence interval 29 to 49
```

This estimate is higher, higher even than the Mh estimate from MARK (which is based on two groups of animals rather than individual differences), and the confidence limits are quite wide, which might be more realistic.

Getting Finished

MARK automatically saves all the results of analyses when you run them in files with the .dbf and .fpt extensions. You do not need to save results manually. (If you copy these files to a new location, be sure to copy *both* of them.)

To exit MARK, select File > Exit or press Alt+F4 or click on the ⊠ button at the top right of the window. To re-open the project, use File > Open and select the appropriate .dbf file. If you do this when a project is already open, MARK will automatically close it. You cannot have two projects open at the same time.

Analyzing Your Experimental Data

Get out the results of your card experiment from the first section of this chapter (page 128). Start a text editor such as Notepad, and enter the capture histories. On each line put one capture history with no spaces between the '0's and '1's, then a space, then the number of cards with that capture history. My results looked like the box below:

```
/* My card experiment */
11101 3;
11100 2;
11001 1;
11000 1;
10111 1;
10101 3;
10100 4;
10011 1;
10001 2;
10000 2;
01100 2;
01010 1;
01001 3;
01000 5;
00111 2;
00110 1;
00101 1;
00100 3;
00011 1;
00010 1;
00001 2;
```

Save the file, giving it a name ending with "inp."

Analyze the data in the same way that you did for the tiger data. You may find that some of the models don't work properly with your data set – look out for $p = 0$ or $p = 1$ in the output. In this case we know the right answers, so we can assess how successful the mark-recapture method and MARK analysis were.

Expected Products

- A presentation (in written, verbal, or presentation form as your instructor prefers) of the results of your card experiment. State which model is correct and compare with the model selection results in MARK and CAPTURE. Also compare the estimates and confidence limits from MARK and CAPTURE with the actual number of cards in the pack.
- A short summary of the results for the analysis for the tiger data from Kanha National Park. We got different results for different models, and you should explain which you think is the best to use to estimate the tiger population for the park. The surveyed area is estimated at 282 km²; calculate the density of tigers (per 100 km²), with confidence limits. The area of the core of the National Park is 940 km²; assuming the density is the same in the rest of the core as it is in the sampled area, what is the total number of tigers?
- Responses in a form indicated by your instructor to the Discussion questions below.

Discussion

1 Compare the results of the card experiment with other people in the class. (If you are working on your own, repeat the experiment at least one more time and compare the results.) How often did the software come up with the right model? Were the estimates generally close to the right answer? Was the right answer always within the confidence interval?

2 Tiger densities for other sites studied using the same field methods and the Mh analysis in CAPTURE are shown in Table 14.2. Compare the results you got for Kanha with each of these 10 sites: Is the tiger density at Kanha higher or lower? Could this difference be due to chance? What criteria would you use to decide?

Making It Happen

Capturing live vertebrates is not something to undertake lightly. Precautions to protect you and the animals from injury and infection are important. Even if done properly, capture is stressful for wild animals. You may need permits from government agencies to trap animals legally in your locality, and permits may also be needed to band or tag animals. So it is only worthwhile if you have a serious research question in mind; doing it *just* to try out the techniques or to generate numbers for analysis is difficult to justify.

Table 14.2 Tiger densities for other sites studied using the same field methods and the Mh analysis in CAPTURE.

Reserve	Tiger density
Tadoba	3.27 (0.59)
Bhadra	3.42 (0.84)
Pench-MP	4.94 (1.37)
Melghat	6.67 (1.85)
Panna	6.94 (3.23)
Pench-MR	7.29 (2.54)
Ranthambore	11.46 (4.20)
Nagarahole	11.92 (1.71)
Bandipur	11.97 (3.71)
Kaziranga	16.76 (2.96)

Monitoring the number of a common animal such as squirrels on your campus might be a useful project. Data from the first year alone would be of limited use, but if followed up by future classes the result would be a time series which could be compared with records of severe weather events, changes in habitat, or food availability.

Capturing invertebrates is less onerous. Insects can be marked with spots of colored paint (a different color for each trapping session), but flying insects often travel over large areas, so it's not clear what area you are sampling.

Perhaps the most original idea for a non-biological mark-recapture experiment is Carothers' (1973) study of taxicabs in Edinburgh, UK. Many cities have a "closed population" of easily recognizable cabs or buses, already individually marked by their license plates.

Further Resources

A description of the study producing the tiger data are in Karanth and Nichols (1998), with more analysis and discussion in Karanth et al. (2004).

The best resource if you want to learn more about MARK, the range of analyses it can do, and the math behind the software is Cooch and White (2006), which is available for free download. Classical references on mark–recapture techniques are those by Otis *et al.* (1978) and Seber (1982). The book edited by Amstrup et al. (2005) is a mine of information on current aspects of mark-recapture studies. Mark-recapture is a well-worn technique and there are several chapters on it in Krebs (1999) and Williams et al. (2002).

PART 4

Species

CHAPTER 15

Estimating "Biodiversity": Indices, Effort, and Inference

Michael E. Meredith

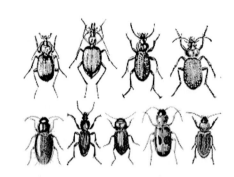

Estimating biodiversity at particular sites and contrasting it among sites is a fundamental undertaking of conservation biologists, especially those working in the field doing inventories and prioritizing sites for protection. We have already touched upon this topic: in chapter 1, "What is biodiversity?", where you looked at samples of spiders from different communities and used the results to explore species richness, species diversity, and the similarity of communities. With samples of 50 individuals, it was easy to do this manually. But most real studies involve much larger numbers of individuals and large numbers of species. For example, the study by Longino et al. (2002) of ants in lowland rain forest at La Selva Biological Station, Costa Rica, involved 7 904 observations of 437 species. Collecting and identifying the specimens required a huge effort spread over several years, but the data analysis was relatively fast, and was, of course, computerized.

Apart from keeping track of the data and adding up the number of species we have collected, sophisticated software lets us:

- estimate the completeness of our list of species and the species richness of the site
- compare the species richness of different sites, even though different numbers of individuals were collected at each
- check estimates of diversity indices to see if they are affected by incomplete sampling
- explore the similarity of communities at different sites or at different times.

Several software packages are available to analyze these kinds of data sets, but here we will introduce "EstimateS", which is both freely available and one of the most widely used packages. Developing a facility with EstimateS (or similar software) is a requisite for learning how to manage the data side of biodiversity inventory work.

Objective

- To build on the concepts introduced in chapter 1, and to explore how they can be implemented and extended with the software package EstimateS.

Procedures

Defining the Community

To use species richness in practice, we need first to define the group of individual organisms to be included. We usually limit it to a particular taxonomic group (e.g. spiders or birds) and a particular place (e.g. La Selva or Yellowstone National Park). Often we specify a trophic level (e.g. insectivores) or a guild (e.g. understorey gleaners).

Richness estimation means counting the number of species, treating all species equally, whether they are endangered endemics or invasive weeds, top predators or primary producers. This is only reasonable if the population is defined so that species are actually reasonably similar. A good question to ask yourself is: "If one species dropped out, would the other species be able to fill the gap?" If the answer is clearly "No," then you need to change the community definition, or abandon the idea of just counting species.

We can rarely study all the individual organisms in our target community, and we usually have to work with samples. Invariably there will be population boundaries defined by our sampling method: for the spiders in chapter 1, specimens were collected by striking tree branches with a stick; you could not compare the results from these collections with an additional site where spiders had been collected from the leaf litter. Results can only be compared if the same sampling methods were used in gathering the data.

Some of the "rare" species we encounter are actually "edge species," on the boundary of our defined groups of organisms. For example, if we were using traps baited with fruit to study frugivorous butterflies in a forest, we might occasionally capture a butterfly from the adjacent prairie (a **spatial edge species**), a nocturnal moth (a **temporal edge species**), or a butterfly which does not feed on fruit (a **methodological edge species**). Deciding which to include can be difficult, as it implies knowing what species *ought* to be present!

Bat Data from Loagan Bunut

The example data we'll use for the first part of this chapter are for bats caught in harp traps in peat swamp forest in Loagan Bunut National Park in Sarawak, Malaysia.

A harp trap has an aluminum frame with four parallel sets of vertical fishing lines (see Figure 15.1). A bat that flies into a harp trap will get caught in the fishing lines and slide down, uninjured, to the bag underneath. The bats are identified, weighed, measured, and released at the place they were captured at the end of the trapping session. Harp traps are very efficient in capturing echo-locating bats, typically forest interior species, which are unable to detect the thin lines of the harp trap.

Download the file LBunut_bats_input.txt from www.esf.edu/efb/gibbs/solving/. Open it in MS Excel®: either right-click on the file name and select Open With > MS Excel, or drag-and-drop the file into an open MS Excel window. (If you use File > Open in MS Excel, the import wizard will start, which you don't need.)

The data file has a description of the data on the first line. On the second line is the total number of species (22) followed by the number of samples (50). Then come the records of bats captured, with a row for each species and a column for each sample. Each sample corresponds to the bats caught on one night at one

Fig. 15.1 A harp trap (aluminum frame with four parallel sets of vertical fishing lines) used to catch bats in peat swamp forest in Loagan Bunut National Park, Sarawak, Malaysia. (Drawing by Jason Hon.)

location. For example, the first column shows that on the first night 6 species were caught (the other 16 rows contain 0), with one species represented by 2 bats and the other 5 species represented by 1 bat each. The second sample had just 2 bats, both from the same species. (LBunut_bats_input.txt is formatted for input into EstimateS. Note that EstimateS does not use species names or sample identifiers.)

Getting Started with EstimateS

Go to the EstimateS web site (http://purl.oclc.org/EstimateS), and fill in the registration form. Then download the installer for the latest version of the software, currently EstimateS 8.0.0 in SetupEstimateSWin800.exe. As with any .exe file, it's wise to save it on your hard disk, run a virus check with an up-to-date virus scanner, and create a Windows Restore Point before running the setup program. Run SetupEstimateSWin800.exe and follow the on-screen instructions. If you use the results from EstimateS in any report or paper, please cite it as:

> Colwell, R. K. 2006. *EstimateS: Statistical estimation of species richness and shared species from samples*. Version 8.0.0. User's Guide and application published at: http://purl.oclc.org/estimates.

Open EstimateS and click OK to agree not to distribute EstimateS commercially. In EstimateS, click on File > Load Data Input File (or press Ctrl+I). In the dialog box, navigate to the file LBunut_bats_input.txt and click Open. A confirmation screen appears with the title of the data set and the number of species (22) and samples (50). There's also a list of optional items: don't worry about these, as we can set options later. Then a dialog appears asking for the format of the input file. "LBunut_bats_input.txt" is Format 1 and there are no rows with sample information or columns with species names.

You should then see a box telling you that the data have been loaded successfully. Go to Diversity > Diversity Settings... (or press Ctrl+T) and then to the Estimators tab. In the top section, click on the radio-button next to "Use classic formula for Chao 1 & Chao 2." Then go to the Other Options tab and check the box next to "Compute Fisher's alpha, Shannon, & Simpson indexes". Leave the other settings as they are and click on Compute. If a box appears asking if it's OK to erase some old Diversity Statistics, click on OK.

A box pops up briefly with a progress bar, then a table of results appears. EstimateS produces a huge mass of figures, far too many to digest. They make much more sense if we use them to draw graphs, so export them from EstimateS and load into MS Excel® (or other spreadsheet software, such as the OpenOffice software (http://www.openoffice.org/), with a good graphics facility).

Click on the Export button at the bottom of the results window, or go to Diversity > Export Diversity Stats. Give the results file a suitable name, such as LBunut_bats_results.txt. Find the file LBunut_bats_results.txt in My Computer and either right-click and select Open With > MS Excel or drag and drop into an MS Excel window.

The results should now appear in the spreadsheet. Plotting will be simpler if you delete the first few lines, so that the column headings appear in Row 1, then save it in MS Excel's .xls format: highlight the first three rows of the spreadsheet and select Edit > Delete... then Entire Row. The column headings should now be in Row 1.

Look at the tab for this spreadsheet near the bottom of the window. It has the same name as the .txt file we imported, LBunut_bats_results, which is too long for some of the procedures we want to use later. Right-click on the tab name, select Rename, and change the name to just "Lbunut." Go to File > Save As..., change Save as Type, at the bottom of the dialog box, to Microsoft MS Excel Workbook (*.xls).

Species Accumulation Curves

When we looked at the input data, we saw that the number of species observed (**Sobs**) was 22. The first question is: "Have we captured all the species in the population?"

In chapter 1, you used a **collectors curve** to answer this question, plotting the number of species versus number of individuals as more and more spiders were added to the sample. The solid line in Figure 15.2 is the species accumulation curve

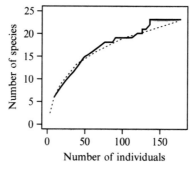

Fig. 15.2 An example species accumulation curve. (See text for details.)

for Loagan Bunut bats, drawn in the same way as we collected more and more samples. In EstimateS, this kind of curve is called a **Species Accumulation Curve**.

This curve is quite jagged, and the shape depends on the order of the samples. But if the samples are properly independent, the order of collection doesn't matter. We can remove the jaggedness by shuffling the order of the samples many times and averaging the curves obtained, to produce the dotted line in the graph. EstimateS has already done this, shuffling the samples 50 times and taking the mean; the results are in column G [Sobs Mean (runs)] of the spreadsheet. Let's plot it.

In the MS Excel spreadsheet, highlight column B and column G:

- Individuals (computed)
- Sobs Mean (runs)

(Click on the B at the top of column B, then hold down the Ctrl key while clicking on the G at the top of column G.)

Start the Chart Wizard by clicking on the toolbar button or using Insert > Chart...

Step 1 : Under Chart Type: select XY (Scatter), then in Chart Subtype choose one with "lines without markers". Click Next.

Step 2 : Check that one curve is displayed with the *x*-axis running from 0 to 200. On the Series tab, you should have one series (Sobs Mean (runs)), with Column B (Individuals) used as the *x* values. If all is in order, press Next.

Step 3 : In the Titles tab, put in suitable names, such as:

- Chart title: species accumulation curve
- Value (*x*) axis: No. of Individuals
- Value (*y*) axis: No. of Species

Press Next.

Step 4 : Select As New Sheet, and name it Species Accumulation Curve; click Finish. Save your work (Ctrl+S).

The curve is not exactly jagged, but it isn't really smooth either. More shuffles of the samples would be needed to make it properly smooth. But EstimateS has another way to do this, using an algorithm developed by Mao Chang Xuan to calculate the values we would get for an *infinite* number of randomizations: this is in column C [Sobs (Mao Tau)]. His algorithms also allow us to calculate the confidence interval and those values are in columns D and E. Add these to the graph as follows.

Go to the Species Accumulation Curve graph and select Chart > Add Data... Click on the small 🔣 icon at the right-hand end of the Range box. Now go to the LBunut results spreadsheet and highlight columns B through E:

- Individuals (computed)
- Sobs (Mao Tau)
- Sobs 95% CI Lower and Upper Bounds.

(Click on the B at the top of column B, then hold down the Shift key will clicking on the E at the top of column E.)

Click on the 🔲 icon at the right-hand end of the Add Data – Range dialog box. Back in the Add Data dialog box, click OK. The Paste Special dialog box opens.

- We want to "Add cells as" **New series**, with "Values (Y) in" **columns.**
- Make sure that the boxes next to Series Names in First Row and Categories (X Values) in First Column are *both* checked.
- Click OK.

Three new lines appear in the graph, the middle one being very close to the original Sobs curve but much smoother. The outer lines are the confidence limits – let's make them dotted lines, the same color as the Mau Tao curve: right-click on the top line and select Format Data Series.... On the Patterns tab, change Style to a dotted line and Color to the same color as the main line (pink). Click OK. Do the same for the bottom line.

The species accumulation curve we have just plotted is clearly still climbing and showing no signs of leveling out at an asymptote. It looks as though our coverage of species is still rather incomplete.

Singletons and Uniques

Singletons are species which are represented by a single *individual* in the collection. If our sampling has been really thorough, we will have caught all the species there twice or more. Are there singletons among the bat species trapped at Loagan Bunut?

Go to the LBunut worksheet in MS Excel. Highlight cell D2 (just below the heading Sobs 95% CI Lower Bound), then go to Window > Freeze Panes. Now scroll down and check the last value in the Singletons Mean column (column H). You'll see there are 8 singletons (out of 22 species in total), so our sampling is not very thorough. Check also the number of **doubletons**: these are species represented by just two individuals in the collection.

If bats of a particular species tend to occur in groups, we'll get a few samples with several bats and lots of samples with none. In this case, the concept of **uniques** – species which occur in only one *sample* – is more appropriate than singletons. **Duplicates** are species which occur in just two samples. Check the number of uniques and duplicates in the Loagan Bunut bat data.

Just as for the species accumulation curve, the *trend* in the numbers of uniques and duplicates (or singletons and doubletons) may be more informative than the final numbers. Let's plot the graphs.In the LBunut spreadsheet in MS Excel, highlight columns B, H, J, L, and N, i.e:

- Individuals (computed)
- Singletons Mean
- Doubletons Mean
- Uniques Mean
- Duplicates Mean.

(Click on the B at the top of column B, then hold down the 'Ctrl' key while clicking on H, J, L and N.)

Start the Chart Wizard by clicking on the toolbar button or using Insert > Chart...

Step 1 : Under Chart Type: select XY (Scatter), then in Chart Subtype choose one with Lines without markers. Click Next.

Step 2 : Check that 4 curves are displayed with the *x*-axis running from 0 to 200. On the Series tab, you should have 4 series, with Column B (Individuals) used as the *x* values. If all is in order, press Next.

Step 3 : In the Titles tab, put in suitable names, such as:

- Chart title: singletons, doubletons, uniques and duplicates
- Value (*x*) axis: No. of Individuals
- Value (*y*) axis: No. of Species.

Press Next.

your work (Ctr+S).

The curves have been smoothed in the same way as the first species accumulation curve we plotted: EstimateS has shuffled the samples 50 times and averaged the results. The shuffling is random, so you will not get exactly the same results each time.

What would you expect to see? When you have only a few samples, you'd expect lots of uniques and singletons; these curves reach a peak at about 30 individuals (8–9 samples) and then start to decline. The duplicates and doubletons reach a peak a bit later, and then also decline. The graph for the Loagan Bunut bats is odd, because the uniques and singletons then start to climb again, and go on climbing. Can you figure out what's happening here?

Estimating True Species Richness

Unless we have collected an enormous amount of data, we will have missed a few rare species. Estimates of species richness based on the number of species observed (Sobs) are generally too low. Can we use the data we have to estimate how many species we have missed and what the true number might be?

Many people have suggested estimation methods, but there is no clearly correct way to estimate the true species richness from a sample. We are trying to *extrapolate* from what we know to a situation for which we have no data. The methods suggested can be grouped as follows:

- Jackknife and bootstrap methods examine the sampling process, asking, "How many species which we know are there would we have missed if we had taken fewer samples, or different samples?" This is then used to estimate the number missing from the actual set of samples.
- Anne Chao's methods consider the number of species which are so rare that they only occur once or twice in the sample, and try to estimate how many are even rarer, so didn't turn up in the sample at all. ACE and ICE (Abundance-based and Incidence-based Coverage Estimators) use the same approach but using species which occur 1–10 times in the sample.
- Finally, we can try to fit a mathematical equation to the species accumulation curve, and use this to predict the number of species where the curve levels off. A favourite is the Michaelis–Menton equation.

More details of these estimators are in the documentation for EstimateS, which you can access on-line at http://purl.oclc.org/EstimateS.

What are the criteria for a good estimator of species richness? If an estimator gives the correct values for the true richness based on the set of samples we have, that value should not change as we add more samples. Plotted with the species accumulation curve, the ideal estimator would be a horizontal line which meets the species accumulation curve where it levels off – like the solid horizontal line in Figure 15.3. In practice, we'd expect the estimate to be very imprecise when we have only observed a small number of individuals, but to settle down to the correct value as we observe more – the dotted line in the graph would be fine. What we often get, however, is an estimate which climbs steadily as we collect more samples, like the dashed line; this leaves us with the same problem – we don't know where it will level off!

EstimateS calculates several estimates of true species richness. Let's plot them and see if any fit our criteria for a good estimator.

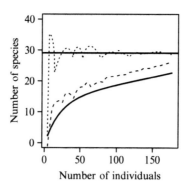

Fig. 15.3 Two contrasting outcomes of attempts to estimate true species richness: an ideal estimator would quickly "level off" (upper line); many estimators continue to climb as more individuals are collected and the end point is uncertain (lower curve). (See text for details.)

Plotting Estimators of Richness

In the LBunut spreadsheet in MS Excel, highlight the following columns:

- B – Individuals (computed)
- C – Sobs (Mao Tau)
- P – ACE Mean
- R – ICE Mean
- T – Chao 1 Mean
- X – Chao 2 Mean
- AB – Jack 1 Mean
- AD – Jack 2 Mean
- AF – Bootstrap Mean
- AI – MMMeans.

Start the Chart Wizard and plot a chart as you did for the singletons, uniques, etc. curve. (If you get a message saying "Your formula contains an invalid external reference...," it may be because the name of the results spreadsheet is too long.) Place the chart in a new spreadsheet and give it a title such as "Species Richness Estimators". Save your work.

The chart is a bit messy with so many lines, and the different colors in the legend are not clear. But if you point to a curve with your mouse, a small box appears with details of the data source. As you'll see, most of the curves are climbing gradually, running parallel to the Sobs curve (the lowest one in the graph). Chao 1 and Chao 2 are climbing more steeply; these are based on the numbers of singletons and doubletons (Chao 1) or uniques and duplicates (Chao 2) and we saw earlier that these are anomalous. The only one which seems to be reasonably horizontal is "MMMeans", the Michaelis–Menton estimator based on the Mao Tau curve for Sobs.

The Michaelis–Menton equation describes the progress of an enzyme-catalyzed chemical reaction. There is no theoretical reason to think it might also describe species accumulation, but it does have the right sort of shape, rising quickly at first and then leveling out. The equation is quite simple:

$$S_{obs} = \frac{S_{max} \times n}{B + n}$$

where S_{max} is the maximum number of species – where the curve levels off, n is the number of individuals in the set of samples so far, and B is the number of individuals needed to get half the maximum number of species, i.e. when $n = B$, $S_{obs} = S_{max}/2$. The MMMeans column gives us the estimate of S_{max} as we add more samples to the sample set. With all the available samples, the best-fitting curve has $S_{max} = 24.57$ (look at the last value in the MMMeans column) and $B = 35.5$ approximately (look down the Individuals column in the LBunut worksheet to find the value corresponding to Sobs $= 24.57/2 = 12.28$ species). Let's plot this curve and see for ourselves if it looks like the species accumulation curve:

Insert a new column next to MMMeans (highlight the column to the right of MMMeans, then use Insert > Column) and call it, say, "MM fitted." In the first row of data type

$$= 24.57 * B2/(35.5 + B2)$$

24.57 and 35.5 are the values for S_{max} and B, while the value in cell B2 is the number of individuals, n. Select all the cells in this column from Row 2 to Row 51, then press Ctrl+D to copy the formula to all the cells in the column. Save.

Now plot the MM fitted curve and the Sobs curve – and the MMMeans curve too for good measure: Highlight columns:

- B – Individuals (computed)
- C – Sobs (Mao Tau)
- AI – MMMeans
- AJ – MM fitted

Start the Chart Wizard and plot a chart as before. Place it in a new worksheet with a title such as "MM curve". Save.

The MM fitted and Sobs curves are very close together on the left of the graph, but above about 120 individuals they start to diverge. The MM fitted curve will eventually level off at 24.57 species, but the Sobs curve looks as if it is heading higher. The Michaelis–Menton curve does not appear to be a good fit for the part of the curve we are interested in.

Comparing Species Richness Between Sites

Very often the important question is not "How many species...?" but "Are there more species at... than...", either comparing two sites or the same site at two points in time. Here we can make inferences based on **interpolation**, which is much safer than extrapolation.

Some harp-trapping has been done in peat swamp forests at Maludam National Park, about 450 km from Loagan Bunut. Unlike Loagan Bunut, Maludam was used for timber production before being established as a national park. In Maludam, 81 bats from 11 species were caught. In Loagan Bunut NP we caught 174 bats from 22 species. Does that mean that Loagan Bunut has more species of bats which can be trapped in harp-traps?

We can't compare the species totals – 11 vs 22 – directly, because the number of bats caught in Maludam is less than half the number caught in Loagan Bunut. If we had carried on trapping at Maludam, we would almost certainly have found more species but, as we've seen, it's difficult to estimate how many more. However, we can estimate how many species we would *expect* to find at Loagan Bunut if we only trapped 81 individuals. We use a subset of the data we collected at Loagan Bunut, with a process known as **rarefaction**. This involves taking samples at random from the Loagan Bunut set until we have approximately 81 individuals, and noting how many species we have found. We do this many times, and average the number of species.

In fact, the smoothing process for the species accumulation curve also uses random selections from the samples, and Rarefaction Curve is an alternative name for the smoothed species accumulation curve. So the figures we need are already in the output from EstimateS:

Go to the results spreadsheet for Loagan Bunut, and look down the Individuals (computed) column to find the number nearest to 81. Note the number of species – Sobs (Mao Tau) – and the upper and lower 95% confidence limits. With 23 samples we'd catch on average 80.04 individual bats, which is the nearest value to 81. And with 23 samples we'd record on average 16.55 species. This is more that the 11 species we recorded in Maludam NP with a very similar number of bats, so it appears that Loagan Bunut is indeed richer.

Now look at the 95% confidence intervals for Sobs for 23 samples: from 10.54 to 22.55. The value of 11 lies within the 95% confidence interval, meaning that finding 11 species (or even fewer) in 23 samples from Loagan Bunut is not improbable.

Simpson's Index: An Alternative to Species Richness

Species richness is a seductive concept, but it is almost impossible to measure. We can rely on getting the common species in our sample, it's the rare species which are the problem. Moreover, rare species are less important to the structure and function of an ecosystem that the common species. (If that's not true of the community you have in mind, you may need to revise how you define it.) So why not use a measure which depends mainly on the number of common species and is not changed much by the number of rare species?

Several diversity indices do just that, including Simpson's Reciprocal index, which you met in chapter 1. Most indices are still biased low when calculated from samples that do not include all the species present in the population, but the bias is small. The original form of Simpson's index for samples (Simpson, 1949) was:

$$D = \sum \frac{n_i(n_i - 1)}{N(N - 1)}$$

where n_i is the number of individuals of species i in the sample and N is the total number of individuals in the sample. When the n_i values are all very large, this is equivalent to the simplified equation used in chapter 1; for small values of n_i, the formula above gives unbiased estimates of the value for the whole population, and the reciprocal of this is used by EstimateS.

Go back to the LBunut spreadsheet and look for the column headed Simpson Mean. (Note that EstimateS does not calculate Simpson's index by default, only if you check the box next to Compute Fisher's alpha, Shannon, & Simpson indexes, on the Other Options tab in Diversity > Diversity Settings..., before running the analysis.)

The values shown for Simpson are actually the Reciprocal Simpson index, $1/D$. The result based on all the samples together is just over 11, which we can think of as indicating that there are 11 common species in Loagan Bunut peat swamps.

Do we have a big enough data set to get a good estimate of Simpson's index, or will it continue to creep up as we collect more bats and discover more species? We'll plot Simpson's index in the same way as the other results: in the LBunut spreadsheet, highlight the Individuals (computed) and Simpson Mean columns and plot a graph as we did before for the species accumulation curve and the species richness estimators.

As you will see, the curve rises very quickly to a value above 10 and then soon settles down near the final value of 11. Unlike the species richness estimators, Simpson's index is not much affected by the size of the sample set. EstimateS calculates two other indices: Fisher's Alpha index and Shannon's index. If you have time, plot these, too, and see how they behave.

Estimating Site Similarity

In chapter 1, you used the Jaccard coefficient of community similarity to contrast pairs of sites, calculating the value based on samples of spiders from the different sites. If sampling is incomplete, we will have missed some species in the communities, and we will have underestimated the number of shared species. EstimateS can calculate the Jaccard coefficients and a range of more sophisticated similarity measures which aim to minimize the effect of incomplete sampling.

The harp-trapping techniques used in peat swamp forests in Loagan Bunut and Maludam were also deployed in lowland forest in Loagan Bunut. Table 15.1 below summarizes the results for the three sites.

Download the file "bats_3sites_input.txt", which has these data formatted for input to EstimateS. Open EstimateS and import the data. Now go to the Shared Species menu and select Shared Species Settings. . . . Check the box next to Compute bootstrap SEs. . . , leave the other settings as they are, then press Compute. If a box appears asking if it's OK to erase some old Shared Species Statistics, click on OK. You can export the results to MS Excel if you wish, but there is nothing here which we can display as graphs.

Remember that comparisons are made between pairs of sites, so the three rows of output correspond to:

1 vs 2 : Maludam peat swamp vs Loagan Bunut peat swamp
1 vs 3 : Maludam peat swamp vs Loagan Bunut lowland forest
2 vs 3 : Loagan Bunut peat swamp vs Loagan Bunut lowland forest.

Columns 3 and 4 show the number of species recorded in each of the two sites, and then (column 5) the number of species record in both sites.

Next come estimates of the true number of species for the two sites using the ACE estimator that we saw on page 147. If you look at the LBunut spreadsheet and check the last value in the ACE Mean column you will see it is the same as the value for ACE for Site 2. ACE uses details of the rare species included in the sample to estimate the number of rare species *not* included. The next column, Chao Shared Estimated uses the same approach to estimate the true number of shared species. For more details, you can refer to the EstimateS documentation. It would be possible to plug these three estimates into the equation for the Jaccard coefficient to get an estimate of its true value, but there are better solutions, as we'll see below.

In the next column, the Jaccard Classic result is the same as the Jaccard coefficient you used in chapter 1:

number of shared species, S_{shared}/S_{total}, the total number of species

Table 15.1 Results for harp-trapping techniques at three sites.

Species	1. Maludam peat swamp	2. Loagan Bunut peat swamp	3. Loagan Bunut lowland
Balionycteris maculata	0	4	0
Cynopterus brachyotis	0	0	2
Emballanura alecto	0	1	0
Glischropus tylopus	0	1	1
Hipposideros bicolor	0	1	0
Hipposideros cervinus	7	18	0
Hipposideros cinaraceus	0	0	1
Hipposideros diadema	0	1	3
Hipposideros dyacorum	0	4	16
Hipposideros galeritus	0	1	1
Hipposideros ridleyi	0	6	4
Hipposideros sabanus	0	1	2
Kerivoula hardwickii	4	14	14
Kerivoula intermedia	13	14	21
Kerivoula lenis	4	0	1
Kerivoula minuta	16	6	10
Kerivoula papillosa	12	16	27
Kerivoula pellucida	11	25	26
Macroglossus minimus	1	0	1
Megaderma spasma	1	0	0
Murina cyclotis	0	1	1
Murina rozendaali	0	1	1
Murina suilla	3	2	6
Nycteris javanica	0	3	7
Pipistrellus tenuis	0	0	2
Rhinolophus acuminatus	0	0	1
Rhinolophus arcuatus	0	0	1
Rhinolophus borneensis	0	0	2
Rhinolophus creaghi	0	25	0
Rhinolophus sedulus	0	22	7
Rhinolophus trifoliatus	9	7	5

The Sørensen Classic index is similar to the Jaccard coefficient, but uses the average number of species in the two communities (or samples):

number of shared species, $S_{shared}/S_{average}$, the average number of species

The Jaccard and Sørensen indices give similar results, and there is a simple relationship between them as shown in Figure 15.4. What we say about the Jaccard index in the next few paragraphs applies also to the Sørensen index.

Chao et al. (2005) made two suggestions to get around the problems caused by incomplete sampling. Incomplete sampling affects rare species, and the first line of attack is to give more weight to abundant species than to rare species. Instead of using just the number of shared species, we use the proportion of individuals in each sample belonging to shared species. For sample X,

U_X = total individuals belonging to shared species/total individuals in all species

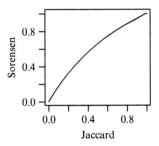

Fig. 15.4 Relationship between the Jaccard and Sørensen indices of community overlap.

and Chao's Abundance-Based Jaccard index is then:

$$U_A U_B / (U_A + U_B - U_A U_B)$$

Fortunately, EstimateS does the calculation for us: see the column headed Chao-Jacc-Raw-Abundance. The values here are higher than Jaccard Classic, showing that it is the relatively abundant species which are shared.

The second step is to estimate the number of rare species missing from the samples, based on details of the species which were found, and to use this to correct the abundance-based index. The values are in the column headed Chao-Jacc-Est-Abundance. They are a little higher, if any, than the "Raw" versions, because the index is in any case not very sensitive to rare species. EstimateS also estimates a Standard Error for the corrected index.

A similar series of indices based on the classic Sørensen index is produced. The equations are somewhat different, but the principles are the same. The last two columns show two well-known abundance-based similarity indices:

- The Morisita-Horn index compares the proportional abundances of the species in each sample, and is given by:

$$\frac{\sum p_{iA} p_{iB}}{\frac{1}{2}(\sum p_{iA}^2 + \sum p_{iB}^2)}$$

where p_{iX} is the proportion of species i in sample X (and is zero if the species is absent).
- The Bray-Curtis index uses the raw numbers:

$$\frac{\sum n_{iMIN}}{\frac{1}{2}(N_A + N_B)}$$

where n_{iMIN} is the smaller of n_{iA} or n_{iB}, the number of individuals of species i in samples A and B (and is zero if the species is absent from either sample), and N_A and N_B are the total number of individuals in the two samples.

Compare the results in the last two columns. Remember that the Maludam sample is about half the size of the others (81 bats vs. 174 and 163) simply because less trapping was done there, and this might affect the results of the Bray-Curtis index.

Finishing Off

EstimateS automatically saves the last set of data to be imported and the last set of results generated, which is useful if something goes wrong. But you should carefully keep the input data files and exported results if you wish to preserve them. Close EstimateS by selecting File > Exit or click on the ⊠ button in the top right corner of the main EstimateS window.

Expected Products

- A presentation (in written, verbal, or presentation form as your instructor prefers) of your conclusions about the species richness of bats in the peat swamp forests in Loagan Bunut National Park. If you think you can give a good estimate of the actual number of species of harp-trappable bats, give your estimates and justify them. If not, explain why none of the values produced by EstimateS seems very good. How can we get better estimates of the actual number of species?
- Use rarefaction to compare the species richness of Loagan Bunut peat swamps with Maludam peat swamps and Loagan Bunut lowland forest (where 163 bats of 25 species were trapped). Calculate the Reciprocal Simpson's index of diversity for the two additional sites, and see if the results agree with the rarefaction analysis. (You can calculate the index from the figures in the table above, or you can download the files Maludam_bats_input.txt and Lowland_bats_input.txt, and use EstimateS.)
- Summarize the results of the analysis of similarity between the three communities of harp-trappable bats. Which measures of similarity do you consider most appropriate for these data? (Hint: the main question is: Which pair of communities is most similar?)
- Responses in a form indicated by your instructor to the Discussion questions below.

Discussion

1 What are possible explanations of the shape of the curves for uniques and singletons for Loagan Bunut peat swamp bat? How would you test your hypotheses?
2 Mist nets were used at Maludam as well as harp traps, and the results are shown below in Table 15.2. How does this information affect the comparison between the sites?

Table 15.2

	Mist netting in Maludam
Balionycteris maculata	3
Cynopterus brachyotis	32
Kerivoula papillosa	1
Kerivoula pellucida	1
Macroglossus minimus	9
Nycteris javanica	2
Penthetor lucasii	1
Rhinolophus trifoliatus	7

3 Select two local conservation issues, one where species richness would be a good criterion for decision-making and one where it would result in a poor decision. In the latter case, what other factors would you need to consider as well as just the number of species?

Further Resources

The documentation of EstimateS explains in detail how the various results are calculated; it is available online at http://purl.oclc.org/EstimateS.

Gotelli and Colwell (2001) give a good review of the issues involved in estimating species richness and using species accumulation curves and O'Hara (2005) looks critically at some of the underlying concepts. The study of ants by Longino et al. (2002) is a good example of the application of these concepts, and Kingstone et al. (2003) have done a major study of bat diversity. Magurran (2004) is a general reference on biodiversity measures, and Hill (1973) describes a framework which shows how several of the common biodiversity indices are related.

The Clear Lake case study (Hunter & Gibbs, 2007, page 30) illustrates the dangers of using the number of species as the sole criterion for management decisions.

CHAPTER 16

Designing a Zoo: *Ex Situ* Centers for Conservation, Research, and Education

Eleanor J. Sterling

Lions, tigers, and bears? Zebras, penguins, and apes? For most people their only encounters with these creatures come at zoos and aquaria, those immensely popular, although very artificial, concentrations of the world's biological diversity. (And of course for orchids, cycads, and palms we have botanical gardens, the plant analogues of zoos and aquaria.) The recreational aspects of these institutions have always been very important, if for no other reason than that almost all of them are dependent on gate receipts and taxpayer support. However, most zoos, aquaria, and gardens have also undertaken an education mission, seeking to teach their visitors about the wonders of life on earth, and often about the fragility of life on earth. In recent years many of these organizations have moved from talking about the problem to undertaking direct action. In particular they have become centers for *ex situ* conservation by breeding species outside their natural habitat. Without zoos many species would be extinct today (Guam rail, Arabian oryx, black-footed ferret, etc.) and it is a sad reality that for some species, captive settings may be the only "habitat" left for them in the future.

Objective

- To learn the basic objectives of *ex situ* conservation programs and to understand the choices that are made when planning species assemblages for these programs.

Procedures

The Scenario

The history of the present-day zoo dates from the late 18th and early 19th centuries, when zoos were established in Vienna, Paris, and London. Worldwide, zoos in the 20th century are focusing increasingly on research and education objectives. In addition, conservation programs, called *ex situ* because they take place outside natural ecosystems (*in situ*), are strongly emphasized in many zoos around the world. In this exercise you will work in teams as scientific advisors in the development of a new, *ex situ* conservation program in Viet Nam. (Note that there are several zoos currently operating in Viet Nam; this exercise does not reflect the status or policies of those zoos.)

Roles

Following the World Zoo Conservation Strategy (IUDZG/CBSG [IUCN/SSC], 1993), the new wildlife center you are advising has chosen as its overall goals:

1 Supporting the conservation of populations of endangered species *in situ* and *ex situ*, and through these conserving natural ecosystems
2 Promoting research to increase scientific knowledge that will benefit conservation of endangered species and ecosystems
3 Increasing education efforts to promote public and political awareness of the necessity for conservation of natural resources.

Each of you will assume a different advisory role and together you must make recommendations for which species to house at the new wildlife center. The different roles for the advisory committee include:

The Curator of Invertebrates: You are particularly interested in the endemic butterflies and moths.

The Curator of Vertebrates: You are interested in Viet Nam's recently discovered mammals, including the saola and the muntjacs, and in rare birds, particularly the pheasants.

The Chair of the Research Department: You are interested in the endemic animals and the recently discovered species because you know it is easier to publish articles about these species. You are also very interested in the herpetofauna, as their numbers are dwindling in the wild due to high levels of trade.

The Chair of the Education Department: You would like to have representatives from as many different Families as possible in order to provide visual examples for your lessons on the variety of life on earth. You are very interested in invertebrates, particularly butterflies, for your program on the stages of life in insects. Finally you want species about which much is known so that you can more easily produce outreach programs.

The Veterinarian: You want the group to choose species that will be easy to maintain with respect to their long-term psychological and physical welfare.

The Field Biologist: In reality, you think that the money spent on this wildlife center should be spent on conservation of species in the wild. Secondarily, you will argue that the wildlife center should focus on well-known species, which have adequate field studies, rather than species that are rare and poorly known.

The Wildlife Center Director: You want to keep the wildlife center financially solvent and will push for species that attract large crowds. You have begun negotiations with a wildlife orphanage in South Africa to bring over a baby giraffe and you are looking for a source for penguins and giant pandas.

Resources

Every animal you recommend must have a role within the structure of the wildlife center's objectives of species preservation, conservation-related research, conservation education, and long-term financial stability. However, there are financial considerations. Your advisory group has 4650 m Vietnamese *dong* available for start-up costs (constructing the exhibits and sleeping quarters, etc.) and 350 m dong for yearly maintenance costs (veterinary care, keeper salaries, animal food, exhibit maintenance, etc.). For the purpose of this exercise, assume that all other costs (outreach programs, administration salaries and costs, etc.) will be met by gate receipts, the government, and donor organizations. As you can imagine, the cost of maintaining large mammals is significantly greater than that of maintaining small mammals, birds, or herpetofauna. These vertebrates, in turn, are much more expensive than most invertebrates. Also, if you recommend species that live in a significantly different environment from Viet Nam (e.g. penguins), the annual maintenance costs will rise proportionately.

Candidate Species

Table 16.1 comprises a list of species from which you could choose to make recommendations to the new wildlife center (though you are by no means limited to these species). We have provided summary information to help you make your choices. Column 1 contains information on each animal's conservation status, derived from the IUCN Red List of Endangered Animals (http://www.iucnredlist.org/). In brief, the categories used to describe a species' status, are: **Extinct** (EX), when there is no reasonable doubt that the last individual in a taxon has died; **Extinct in the Wild** (EW), when a taxon is known only to survive in captivity well outside the past range (A taxon is presumed extinct in the wild when exhaustive surveys in known and/or expected habitat, at appropriate times throughout its historic range have failed to record an individual); **Critically Endangered** (CR), when a taxon is facing an extremely high risk of extinction in the wild in the immediate future; **Endangered** (EN), when a taxon is not Critically Endangered but is facing a very high risk of extinction in the wild in the near future; **Vulnerable** (VU), when a taxon is not Critically Endangered or Endangered but is facing a high risk of extinction in the wild in the medium-term future; **Data Deficient** (DD), when there is inadequate information to make a direct, or indirect, assessment of its risk of extinction based on its distribution and/or population status; and **Lower Risk** (LR), when a taxon has been evaluated and does not satisfy the criteria for any of the categories Critically Endangered, Endangered or Vulnerable. Taxa included in the Lower Risk category are further separated into: **Conservation Dependent** (CD), taxa which are the focus of a continuing conservation program, the cessation of which would result in the taxon qualifying for one of the threatened categories above within a period of five years; Near Threatened, taxa which do not qualify for Conservation Dependent, but are close to qualifying for Vulnerable, and **Least Concern**, taxa which do not qualify for Conservation Dependent or Near Threatened status. "Not listed" refers to species not included in IUCNs Red List of Threatened Animals.

In Table 16.1 there are two cost estimates: **Maintenance Costs** represent estimated annual costs for maintaining a minimum group size (in millions, M, of Vietnamese *dong*, where $1=10,000 dong and 1M dong=$100). Remember that you have a limited amount of money for yearly maintenance and therefore have to pay special attention to this category. **Start-Up Costs** (building cages, obtaining animals, etc.) for the minimum group size (in M of dong). **Range** is the known distribution of the species. **Husbandry** summarizes knowledge of husbandry

Table 16.1

Species	Common name	Red List category	Mainten-ance costs	Start-up costs	Range	Knowledge of Husbandry	Knowledge in Wild	Captive Population (ISIS)	Flagship?	Reintro-duction potential	Additional information
MAMMALS											
BATS											
Pteropus giganteus	Indian Flying Fox	Not listed	5	200	India, Burma, and Sri Lanka	Good	Fair–good	37	N	Good	Important pollinators of the plant genus *Freycinetia*. Make regular mass migrations, following the seasons for fruit ripening.
Rhinolopus ferrum-equinum	Great Horseshoe Bat	Not listed	5	200	Europe, Africa, southern Asia	Fair–good	Fair–good	0	N	Good	Horseshoe bats live in groups and roost in damp, dark places, such as caves.
MARSUPIALS											
Petrogale xanthopus	Yellow-footed Rock-Wallaby	Lower risk: near threatened	5	200	Australia, New South Wales	Fair–good	Good	11	N	Good	Live among rocks, usually near water.
PRIMATES											
Gorilla gorilla gorilla	Western Lowland Gorilla	Endangered/A2cd	70	450	Cameroon, Congo, Gabon	Good	Fair–good	97	Y	Poor-fair	Live mostly in lowland rain forest; form stable 1 male-multifemale family groups.
Lemur catta	Ring-tailed Lemur	Vulnerable/A1c	15	300	Madagascar	Fair–good	Fair–good	180	Y	Fair	Only species in genus. Lives in scrub, spiny desert, dry and gallery forest.
Macaca mulatta	Rhesus Macaque	Lower risk	15	150	Afghanistan and India to Thailand and southern China	Good	Good	21	N	Fair	Live in mult-male and multi-female social groups of female bonded matrilines. Males are peripheral to the group and change troupes every few years.

(Continued)

Table 16.1 (*Continued*)

Species	Common name	Red List category	Maintenance costs	Start-up costs	Range	Knowledge of Husbandry	Knowledge in Wild	Captive Population (ISIS)	Flagship?	Reintroduction potential	Additional information
Nycticebus pygmaeus	Lesser Slow Loris	R/lc	15	200	Viet Nam, Laos, Cambodia and Yunnan	Fair	Poor	30	N	Fair	This animal's nocturnal behavior means that the enclosure should include reverse lighting, which adds an additional expense to initial costs.
Pan troglodytes	Chimpanzee	Endangered/A2cd	60	300	Central and west Africa	Good	Good	135	Y	Poor-fair	Multi-male, multi-female communities with unrelated females and their offspring within a core of related males.
Pongo pygmaeus	Borneo Orangutan	Vulnerable A1cd, C1	60	300	Borneo	Good	Fair	56	Y	Fair	Recent genetic and fingerprinting evidence suggests that there are 2 species of orangutans (this spp. and the Sumatran orang) that diverged 1.5 million years ago.
Pygathrix nemaeus	Red-shanked Douc Langur	Endangered/A1cd	15	150	Viet Nam, Laos,	Fair	Poor	1	Y	Fair	Mostly eats leaves; lives in lowland rain forest endangered in past by war and now by agriculture.
Pygathrix nigripes	Black-shanked Douc Langur	Endangered/A1cd	15	150	Viet Nam, Cambodia	Poor	Poor	1	Y	Poor	Similar in habits and habitat to *P. nemaeus*, partially sympatric with *P. nemaeus*.

Rhinopithecus avunculus	Tonkin Snub-nosed Monkey	Critically endangered/C1, E	15	150	Viet Nam	Poor	Poor	(1)	Y	Poor	Heavily hunted. Estimated 150 individuals left in 1995.
Trachypithecus delacouri	Delacour's Langur	Critically endangered/A1d, C2a	15	150	Viet Nam	Poor–fair	Poor	(1)	Y	Poor	Heavily hunted. Estimated 200 individuals left in 1995.
Trachypithecus francoisi	Francois' Langur	Vulnerable/A1cd+2cd, C2a	15	150	Viet Nam, southern China, Laos	Poor–fair	Poor	(1)	Y	Poor	Heavily hunted.
RODENTS CALLOSCIURUS											
Callosciurus erythraeus	Pallas's Squirrel	R/lc	5	120	S and SE Asia	Fair	Fair–good	1	N	Good	Emits and responds differently to different types of anti-predator and copulatory vocalizations in the wild and in captivity. Thick-furred tree dweller.
Cynomys ludovicianus	Black-tailed Prairie Dog	Not listed	2	75	Prairies and Rocky Mountains of North America	Good	Good	77	N	Fair	Eradicated across much of former range. Grazing important for maintaining grassland ecosystem. Large groups (>60 animals) needed to establish new populations.
Heterocephalus glaber	Naked-mole Rat	Not listed	15	120	Eastern Africa (Ethiopia, Somalia, and Kenya)	Good	Good	15	Y	Poor	Only species in genus. Unusual in that breeding restricted to one female (the "queen") and one to three males.

(Continued)

Table 16.1 (Continued)

Species	Common name	Red List category	Maintenance costs	Start-up costs	Range	Knowledge of Husbandry	Knowledge in Wild	Captive Population (ISIS)	Flagship?	Reintroduction potential	Additional information
Hylopetes alboniger	A flying squirrel with no common name	Endangered/A1c	5	120	S and SE Asia	Poor	Poor	0	N	Poor	Nocturnal. In "flying," flying squirrels leap spread-eagle and use their outstretched gliding membranes for gliding and their bushy tails for guidance.
Typhlomys chapensis	A doormouse with no common name	Critically endangered/ B1+2cd	2	120	Viet Nam	Poor	Poor	0	N	Poor	Fragmented populations threatened by continued habitat degradation.
EVEN TOED UNGULATES											
Bos gaurus	Gaur	Vulnerable/ A1cd+2cd, C1+2a	20	250	S and SE Asia	Fair–good	Poor–fair	20	Y	Poor	Heavily poached. A domesticated form is used both as a work animal and for meat production. A domestic cow has successfully served as a surrogate mother to an implanted gaur embryo.
Bos javanicus	Banteng	Endangered/ A1cd+2cd, C1+2a	20	250	S and SE Asia	Fair–good	Low	17	N	Fair	Hybridizes with domestic cattle. Species thrives best in areas where shifting cultivation is practiced.
Bos sauveli	Kouprey	Critically endangered/ A2d, C1+2a, D1	20	250	SE Asia	Poor	Poor	0	Y	Poor	Heavily poached. Low reproductive rate. Remnant populations

Cervus nippon pseudaxis	Tonkin Sika	Critically endangered/D1	30	250	Viet Nam	Fair	Fair	14	N	Poor	in areas either plagued by civil unrest or conversion to agricultural land. Success with captive breeding of the gaur can be used as a model. Sika deer females introduced to US are capable of becoming pregnant as calves.
Giraffa camelopardalis	Giraffe	Lower risk: conservation dependent	70	900	Africa	Good	Good	74	Y	Fair	Only species in genus. Although not rare, declining as habitat converted to agriculture. Most populations now restricted to national parks.
Hippopotamus amphibius	Hippopotamus	Not listed	70	1,100	Africa	Good	Good	92	Y	Fair	Only species in genus. Some populations declining due to poaching for meat, fat, hide, and ivory. A common species in captivity where it breeds readily; many captive females on birth control.
Pseudoryx nghetinhensis	Saola	Critically Endangered	20	250	Viet Nam	No captive animals	Poor	0	Y	Fair	First described in 1993, this highly elusive species is heavily hunted and at risk because of its small range size.

(Continued)

Table 16.1 (Continued)

Species	Common name	Red List category	Maintenance costs	Start-up costs	Range	Knowledge of Husbandry	Knowledge in Wild	Captive Population (ISIS)	Flagship?	Reintroduction potential	Additional information
ODD TOED UNGULATES											
Dicerorhinus sumatrensis	Sumatran Rhinoceros	Critically endangered/ A1bcd, C2a	70	1,200	SE Asia	Fair	Fair	1	Y	Poor	Smallest, and possibly most endangered, of the rhinos. Numbers have dropped by 50% in last 10 years. Lives in rain forest and mountain moss forest.
Diceros bicornis	Black Rhinoceros	Critically endangered/A1abc	70	1,500	Africa, south of the Sahara	Fair	Fair	26	Y	Fair	High poaching pressure to meet demand for rhino horn from various Arab and Asian nations.
Equus bircbelli	Burchell's Zebra	Not listed	40	750	Rich grasslands over much of southern and eastern Africa	Good	Good	41	Y	Fair	Zebras live in small family groups consisting of a stallion and several mares with their foals.
Tapirus indicus	Malayan Tapir	Vulnerable/ A1c+2c, B2cd+3a, C1+2b	50	400	SE Asia	Good	Poor	53	Y	Fair	Solitary and nocturnal. Main threat is the conversion of lowland forest to plantations and cultivation, but poaching also important.
PROBOSCIDEA											
Elephas maximus	Asian Elephant	Endangered/A1cd	140	900	Asia	Good	Good	103	Y	Fair	Only species in genus. Major problems faced by this species are habitat degradation and fragmentation of populations.

CARNIVORES

Ailuropoda melanoleuca Giant Panda	Endangered/B1, 2c, C2a	20	1,200	Mountain and forest regions Tibet and south-western China	Poor–fair	Fair	5	Y	Fair	Difficult to export from China. Fewer than 1,000 specimens exist in the wild. Feeds almost exclusively on bamboo.
Ailurus fulgens Red Panda	Endangered/C2a	10	150	Nepal to southwestern China	Good	Fair	21	Y	Fair	Only member in genus. Nocturnal and may live alone, in pairs, or in family groups.
Chrotogale owstoni Owston's Palm Civet	Vulnerable/A1cd	10	150	SE Asia	Fair	Poor	3	N	Fair	Probably confined to riverine habitats, areas usually under heavy human pressure for rice cultivation.
Cuon alpinus Dhole	Endangered	10	150	Asia	Fair	Poor	9	N	Fair	This pack-living canid preys on medium-sized livestock important to local herders.
Cynogale bennettii Otter-civet	Endangered/A1ce, C2a	10	150	SE Asia	Poor	Poor	0	Y	Fair	Heavily affected by human colonization and the expanding rice culture.
Felis silvestris Wildcat	Species not listed by IUCN but numbers are on the decline	40	175	North Africa and SW Asia	Fair–good	Fair–good	4	N	Fair	Hybridizes with domestic cats. However, genetically distinct populations do exist. One subspecies listed as Vulnerable.
Lutra canadensis River Otter	Not listed	40	225	North American lakes and streams	Fair–good	Fair–good	77	Y	Fair	Populations threatened by wetland loss and vegetational change. However, for an otter, this species breeds well in captivity.

(Continued)

Table 16.1 (Continued)

Species	Common name	Red List category	Mainten-ance costs	Start-up costs	Range	Knowledge of Husbandry	Knowledge in Wild	Captive Population (ISIS)	Flagship?	Reintro-duction potential	Additional information
Lutra perspicillata	Smooth-coated Otter	Vulnerable/A2cd	10	200	S and SE Asia	Fair	Poor	1	N	Good	Most common of the Asian otters, through most of its range. However, construction of bridges and mining activities along their riverince habitats have caused a decline in their numbers.
Lynx lynx (Felis lynx lynx)	Eurasian Lynx	Not listed by IUCN but populations are on the decline	20	225	Norway, Sweden, and Finland, eastern Europe and western Asia	Good	Fair	25	N	Fair	Most extensive reintroduction effort for any felid species.
Mustela strigidorsa	Back-striped Weasel	Vulnerable/C2a	10	150	Northern SE Asia	Poor	Poor	0	Y	Fair	Information on biology and status of this species is extremely scarce.
Neofelis nebulosa	Clouded Leopard	Vulnerable/A1cd	70	200	S and SE Asia	Good	Good	49	Y	Poor	Poached for its prized pelt used for the manufacture of jackets.
Panthera tigris	Tiger	Endangered/A2cd	120	200	Asia	Good	Good	47	Y	Poor	Now persists only in isolated pockets of its former range due to poaching and habitat degradation.
Phoca vitulina	Harbor Seal	Not listed	80	500	North Atlantic coasts	Good	Good	44	Y	Fair	Ranges widely throughout the North Atlantic Ocean.
Ursus thibetanus	Asiatic Black Bear	Vulnerable/A1cd	70	200	Asia	Fair–good	Poor–fair	44	Y	Fair	Suffers from high demands as pets and for products ranging from gall bladders to paws.

BIRDS

Ardea herodias	Great Blue Heron	Not listed	5	150	Alaska to the Galapagos Islands	Poor	Good	16	N	Fair	Colonial nester. Populations expanding in many parts of range.
Chrysolophus amberstiae	Lady Amherst's Pheasant	Lower risk: near threatened	5	100	China, Burma	Fair	Good	58	Y	Fair	This beautiful ruffed pheasant has been kept for centuries as an ornamental bird and is a guaranteed crowd pleaser.
Ciconia boyciana	Oriental Stork	Endangered/C1	5	100	NE Asia, Japan	Fair–good	Fair	8	N	Fair	World population estimated at 2 500. Threats include poaching and human disturbance, draining of wetlands for agriculture and disease control, burning and cutting of nesting trees, pesticides and other pollution.
Crocias langbianis	Grey-crowned Crocias	Endangered	5	100	Viet Nam	Poor	Poor	0	N	Fair	Extremely little information is available on this recently re-discovered species.
Leptoptilos dubius	Greater Adjutant Stork	Endangered/C1	10	100	S and SE Asia	Poor	Poor	5	N	Poor	Now restrictred to a few small colonies and scattered pairs; mixed colonies formerly of "millions" of birds now the largest known has 31 nests.

(Continued)

Table 16.1 (Continued)

Species	Common name	Red List category	Maintenance costs	Start-up costs	Range	Knowledge of Husbandry	Knowledge in Wild	Captive Population (ISIS)	Flagship?	Reintroduction potential	Additional information
Leptoptilos javanicus	Lesser Adjutant Stork	Vulnerable/C1	10	100	S and SE Asia	Poor	Poor	6	N	Poor	World population now below 10,000 and still declining. Extinct in several of its range states as a result of habitat loss (both the cutting of nesting trees and the draining of feeding areas), hunting and human disturbance.
Lophura edwardsi	Edward's Pheasant	Critically endangered/ B1+2c, C1+2a, D1	5	100	Viet Nam	Fair	Poor	34	Y	Poor	Thought to be extinct in the wild until late 1996, when several birds were collected from Viet Nam.
Lophura imperialis	Imperial Pheasant	Critically endangered/ C1+2b	5	100	Viet Nam	Poor	Poor	1	Y	Poor	Very rare and elusive bird. Hunted for food and feathers. Appears to be a hybrid
Lophura hatinhensis	Vietnamese Pheasant	Endangered/ B1+2c, C1	5	100	Viet Nam	Fair	Poor	3	Y	Poor	Common in areas with relatively undisturbed, closed canopy forest; areas under high pressure to be logged and developed. Species status unclear; maybe inbred population of edwardsi
Phoenicopterus roseus	Greater Flamingo	Not listed	10	150	Asia, Africa	Good	Good	56	Y	Fair	Generally secure populations live in association with large brackish lakes.
Pygoscelis adeliae	Adelie Penguini	Not listed	300	2,000	Antarctica	Good	Good	2	Y	Poor	Climate change has resulted in increased snow in parts of Antarctica, and may be responsible for declines in populations.

AMPLIBIANS AND REPTILES

Batagur baska	River Terrapin	Critically Endangered	1	50	S and SE Asia	Fair	Poor	5	Y	Fair	Harvested for meat in India.
Boa constrictor	Common Boa	Not listed	5	100	Central and South America	Good	Good	123	N	Fair	Occupies a variety of habitats, including heavily disturbed ones.
Chamaeleo camaeleon	Common Chameleon	Not listed	10	50	North Africa, South Europe, India	Good	Good	5	Y	Fair	This primarily tree-dwelling lizard can change body color to camouflage itself.
Crocodylus niloticus	Nile Crocodile	Not listed	20	300	Africa	Good	Good	28	N	Fair	Although still common, due to hunting this species rarely exceeds 6 m whereas it formerly attained a length of almost 9 m.
Crocodylus siamensis	Siamese Crocodile	Critically endangered/A1ac	20	200	SE Asia	Good	Poor	16	N	Poor	Uncertainties regarding taxonomic status of different populations complicate conservation efforts. Threatened by habitat loss and skin-hunters.
Cuora trifasciata	Chinese Three-striped Box Turtle	Critically Endangered	1	50	Southern China and SE Asia	Poor	Poor–fair	9	Y	Poor	Data deficient.
Dendrobates tricolor	Poison Dart Frog	Not listed	0.5	70	Ecuador	Good	Good	4	Y	Fair	Quite beautiful– guaranteed to be a crowd pleaser. Skin secretes a substance that is toxic to birds and small animals and is used by native S. Americans to coat their arrow tips.

(Continued)

Table 16.1   (Continued)

Species	Common name	Red List category	Maintenance costs	Start-up costs	Range	Knowledge of Husbandry	Knowledge in Wild	Captive Population (ISIS)	Flagship?	Reintroduction potential	Additional information
Naja naja	Asiatic Cobra	Not listed, but listed on CITES Appendix 2	2	50	India and Central Asia	Fair	Good	6	N	Fair	This 1.7-m snake kills several thousand people every year, mostly because it visits houses at twilight to catch rats.
Limnonectes blythii	Blyth's wart Frog	Near Threatened	0.5	50	SE Asia	Poor	Poor–fair	0	N	Fair	Data deficient.
Terrapene carolina	Eastern Box Turtle	Lower risk/near threatened	1	50	Mexico, US	Good	Good	14	N	Fair	Found in wooded areas from the eastern US.
INSECTS											
Apis dorsattta	Giant Honeybee	Not listed	0.25	10	India, Indonesia and central China	Good	Fair–good	No data	N	Good	Builds combs nearly three meters in diameter.
Apis mellifera	European Domestic Honeybee	Not listed	0.25	10	Widespread	Good	Good	No data	N	Good	Critical pollinator for wild plants and agricultural crops; familiar to many people.
Basilarchia (Limentis) archippus	Viceroy Butterfly	Not listed	0.25	10	Neotropics	Good	Good	No data	N	Good	Mimic of the monarch butterfly, which is distasteful to predators.
Carausius morosus	A stick-insect	Not listed	0.25	10	Southeast Asia, Europe	Good	Good	No data	N	Good	Extensively used in studies on insect physiology.
Graphium (Pathysa) phidias	A swallowtail butterfly	Recorded from small area so possibly Vulnerable	0.25	10	Viet Nam, Laos	Good	Poor	No data	N	Good	Little information is available on the biology or conservation of this species.
Hamadryas arinome anomala	Cracker butterfly	Not listed	0.25	10	Neotropics	Good	Good	No data	N	Good	An attractive butterfly abundant throughout its range.

Species	Common name	Conservation status			Distribution						Notes
Libellula forensis	A common dragonfly	Not listed	0.25	10	Widespread	Good	Good	No data	N	Good	Important food source for larger aquatic animals such as fish and useful water quality indicator.
Malacosoma californicum	Tent Caterpillar	Not listed	0.25	10	US	Good	Good	No data	N	Good	Common throughout range. Easily reared in captivity.
Oecophylla smaragdina	Green ant	Not listed	0.25	10	Southern Asia, Australia, Africa	Good	Good	No data	N	Good	Important predator of insect pests of cashew and tea plantations in Australia.
Somatochlora hineana	Hine's Emerald Dragonfly	Endangered/US Endangered Species Act	0.25	10	Patches in Illinois and Wisconsin, US	Good	Poor	No data	N	Fair	Most of the wetland habitat that this dragonfly depends on for survival has been drained and filled.
Stilpnochlora incisa	Peruvian Bush-cricket	Not listed	0.25	10	South America	Good	Fair	No data	N	Good	Crickets play an important role in myth and superstition, and are equated with good fortune and intelligence.
Teinopalpus aureus	Golden Kaiser-I-Hind (a butterfly)	Data deficient	0.25	10	China, Viet Nam	Good	Poor	No data	N	Good	Little information is available on the biology of this species.
Tenodera aridifolia sinensis	Chinese Praying Mantis	Not listed	0.25	10	Native to eastern Asia but have been introduced to North America	Good	Fair-good	No data	N	Good	Now the largest mantid in North America. Ranges from 7 to 10 cm in length.

information in captivity (poor, fair, good), whereas **Knowledge in Wild** is knowledge of ecology and behavior in the wild (poor, fair, good). **Captive Population** lists International Species Information Systems (ISIS) data on current worldwide institutions holding captive populations for 1997, though these are undoubtedly low in many cases. Numbers in parentheses signify data not from ISIS. **Flagship** lists whether the species can be considered a popular "flagship" species (yes = Y, no = N). **Reintroduction Potential** summarizes the potential for reintroduction in native habitat (poor, fair, good). **Additional Information** includes some other data to consider.

Guidelines

In addition to keeping in mind the zoo's objectives and financial parameters (expenses and gate receipts), you should ask yourself several questions when selecting the species for the Wildlife Center program:

- Which species are most in need of captive-breeding programs? Are they immediately threatened *in situ*?
- To what extent will you emphasize conservation status of the species, given that endangered or rare species are most important to your objectives, but common species are often easier to maintain successfully in captivity?
- Should you emphasize native or non-native species (the latter may be more likely to attract visitor attention)?
- Which species do we know enough about to manage a successful captive-breeding program (e.g., those with prior successful captive-breeding programs in other zoos, or those with relatively comprehensive information about life history and behavior from captivity or the wild)?
- Are any of the species you are considering taxonomically unique? If so, are they the only member of their Family? Genus?
- Think about whether or not there are "flagship" species that you could include in the program. These species might be of interest to the public because they are "charismatic" and can therefore draw attention to an endangered ecosystem.

One final thought in choosing species to recommend for maintenance in the new wildlife center: consider the potential for reintroducing or restocking these animals in the wild. What is threatening the species in the wild and can these threats be controlled enough to warrant reintroduction/restocking? What are the cultural or behavioral characteristics of the species that will have to be maintained or taught in the captive populations to ensure survival in the wild? What are some of the medical constraints on putting captive members of these species back in the wild?

Expected Products

- Prepare a report (written or verbal) for the new wildlife center commission, listing species you would recommend them to maintain in their program. Justify your recommendations based on the objectives of the wildlife center and on financial parameters.
- Responses in a form indicated by your instructor to the Discussion questions below.

Discussion

1 Would the money you spent on maintenance of animals in captivity be better spent on maintaining organisms in the wild?
2 What has the success rate been for reintroducing/re-stocking animals in the wild? Which groups of species have been most successful? Why? Should *ex situ* programs include species for which there is no current hope for reintroduction? Why or why not?
3 If skins or other DNA-containing specimens exist for extinct species, should *ex situ* programs invest in cloning these individuals and reconstituting a species? Why or why not?

Making It Happen

Obviously, you would not just take a week to decide what animals would be included in a new zoo. The process would be much more time-consuming and exhaustive. However, many zoos are now reconsidering the status and composition of their collections. In the past, choice of which animals to include in the collections was often made haphazardly, or through personal choice. Zoos are realizing that they do not have enough space to responsibly care for all the species in their collections and they are trying to reevaluate the collection policies. The information that you have gathered is but a part of the process of decision-making that the zoos will go through to organize their future investment in collection management.

You might want to visit your local zoo and use the criteria listed in this exercise to analyze the existing collections. Does there seem to be a system for species acquisitions? Do they emphasize endangered or taxonomically-unique species? Do they use species to highlight conservation issues for *in situ* habitat? Do they mostly have large mammal species or do they also include invertebrate exhibits? You might see if the zoo has a volunteer program where you could help with education or research on endangered species. If you are interested in invertebrates, you could offer to help put together an exhibit on butterfly life stages or the importance of pollinators to biodiversity and agriculture.

Further Resources

Key resources on zoo history and operations and the role of zoos in society are Kisling (2001) and Hanson (2002). Norton et al. (1996) is an intriguing look at the ethics of zoos.

CHAPTER 17

Plant Reintroductions: Reestablishing Extirpated Populations

Richard B. Primack and Brian Drayton

Simply buying conservation land and then guarding it against external threats represents only part of an overall conservation strategy. Isolated conservation areas lose species over time due to a combination of human impact, random environmental events, and population processes. The species that are lost tend to be the ones with the greatest conservation value. For example, the Middlesex Fells, a conservation area in Massachusetts, has lost one-third of its native plant species over the last 100 years, including most of its orchids and lobeliads. Even when such areas are carefully monitored and protected, the lost species are often not able to return because of a fragmented landscape. Conservation biologists working with plants and animals are increasingly attempting to reintroduce species into impoverished sites within their historical range. The assumption is that many plant species are not capable of dispersing back to the original sites on their own and that biologists have to assist in the dispersal process. However, this field of plant reintroduction is still in its infancy, and there are few documented examples of success, as defined as a reproducing population that is stable or expanding in numbers. Active investigations are underway to determine the most effective ways to carry out reintroductions. Is it better to start new populations from seeds, seedlings, or adult plants? How many seeds, seedlings, or adults need to be used? Do the sites need to be modified before planting to increase the possibility of success? Do plants need to be taken care of after being planted (a "soft release"), for example, by being watered, being fertilized, or having competing species removed?

Objective

- To design and implement a reintroduction plan for a plant species and to determine the effectiveness of reintroduction as a tool to establish new populations of plant species.

Procedures

Initial Considerations

This plant-based project provides a practical introduction to many fundamental questions that arise in all reintroduction and restoration work, whether of plants or other organisms. Plants are generally easier to work with than animals at many points in the reintroduction process. However, if your particular interest is in animal species, you should evaluate how the project would have to be different if you were working with an animal species.

Choosing a Target Site and Target Species

For this exercise, you need to identify two field sites. First, you should choose a site that appears to be recovering from past human activities as the target site for your experiment. These habitats might include heavily logged forests, degraded wetlands, over-used dune systems, isolated city parks, vacant lots, overgrazed rangelands, abandoned mines, or old industrial sites.

Second, you should choose a less-disturbed reference site that appears to be similar in terms of such variables as soil type, soil moisture, the amounts of herbaceous plant cover and woody vegetation, and the degree of shading. A comparison of the target site with this nearby reference site could reveal one or more native species that could potentially grow at the target site, but are currently absent. Human activity may be responsible for the loss of species from this site. What can you do to get those missing native species to grow at the target site?

Designing the Reintroduction Plan

Your basic task is to determine the best possible way of creating a new population (or several populations) of a plant species on your target site. You need to design an experiment that will yield useful information. Be aware as you think about this that although the data from your projects first year will be interesting and important, it will probably be a multi-year project. This means that others will continue your work at some point, and you should bear these future collaborators in mind when keeping records and in designing the monitoring and maintenance of sites. Experiments that compare effects at more than one site, or from more than one method, often yield valuable results. A comparison might be made between using seeds versus adults (assuming both are available); planting the plants into the intact ground versus digging up the ground with a spade to reduce plant competition; or leaving the plants on their own after transplanting ("hard release") versus watering, fertilizing, or weeding out competing plants ("soft release"). Other questions of current research interest include: How many seeds do you need to create a new population? How many plants do you hope to have growing at the target site? How long should the monitoring continue, and at what frequency?

In the Field

Material for the reintroduction, whether by seed or transplanting, should be collected from an area as close to the target site as possible, preferably the reference site. If the species is a perennial, you may be able to collect some seedlings or adult plants and transplant them to the target site. Good horticulture is the key to success, in particular, not damaging the root system or other plant parts in the process. The best method might be to dig up a whole clump of soil around and including the plant, and then place this into a hole dug at the target site. Discussions with people who have gardening experience may be helpful. Such transplanting is often best done early or late in the growing season, or during periods of wet weather. It is important not to remove so many plants from the reference site that the original population is damaged; as a rule of thumb, probably less than 20% of the seeds should be removed, or less than 10% of the plants dug up.

Reintroductions should be made into clearly marked plots. If the plots are square (perhaps 0.5 m or 1 m on a side), then all four corners could be marked with stakes. If the plots are round (perhaps 1 m in diameter), then only the center point has to be identified with a stake. You should consider that the stakes may be removed or knocked down; exactly measure the distance and compass heading from each stake to permanent features of the landscape (e.g. "From the central stake in plot #3 to the isolated magnolia tree, the distance is 3.4 m in a direction of 83°"). For the vast majority of species, it is not known how to create a new population. A comparison of various methods will give you information on how to carry out a successful reintroduction, and may well provide some new information about the biology of the target species.

Record-keeping

In order to evaluate the success of the experiments, careful records will have to be taken on how many seeds, seedlings, or adult plants were used, what procedures were followed, and the exact location of the plots that received plants. Ideally, you should also note the dates of experimental work, the dates of monitoring visits, and the date and general location of sources for your material (where you collected seeds or plants for transplant). These records will form the basis of a monitoring program over the following growing season and the coming years. All of this information will aid in the interpretation of results. Your records should include a clear description of the reintroduction plan as a whole, including the criteria used to choose species, the rationale for the size of the experiment, the desired size of restored populations, and criteria by which success will be measured.

Box 17.1   Equipment

- Local field guide to identify plant species
- Notebook and pencils
- Wooden or metal stakes cut to 30–40 cm lengths
- Compass
- Tape measure
- Plastic bags, ties, and labels for collecting seeds
- Trays, boxes, or large bags for transplanting adult plants
- Shovels or trowels
- Buckets and bottles for watering plants.

Records of your activities need to be exceptionally clear and then written up as progress reports. These materials should be photocopied and stored in electronic format and kept in several secure, well-known places so that future classes can follow-up on these experiments. How many seeds were used? Where were the plants obtained? Where exactly are the plots located? What treatments were done to the plots and when? You may wish to verify that your records are intelligible to others besides the writers, asking classmates to examine them, and to try to locate field sites.

Photographic records of reference site, target area, and experimental populations can be very helpful as reminders and illustrations about the sites and the procedures you undertake, and can provide useful data in their own right.

A Final Comment to Future Classes Doing the Reevaluations

It is often difficult to create new populations of plant species. If no plants can be found in the experimental plots after one or two growing seasons, consider repeating the experiment (because some years may be better than others for reintroductions), or try using plots in different areas, more plots, or greater numbers of seeds and adult plants. Consider removing more of the competing vegetation, adding additional soil or providing shading, depending on the circumstances. Something is bound to work.

Expected Products

- Rationale for choosing target site and target species
- Formal, detailed reintroduction plan
- Documentation of transplant activities sufficient to permit assessment of transplant success in future years
- Assessment of what worked and what didn't with recommendations for follow-up transplant efforts (for resurvey years)
- Responses in a form indicated by your instructor to the Discussion questions below.

Discussion

1 What if the classes visiting the sites in future years are unable to detect living plants on any plots? Does it just mean that these techniques did not work at this particular time and place? How could the chance of success be increased?
2 What if the follow-up in one or two years shows numerous plants surviving, and even flowering? Does this mean that the reintroduction was a success? Why or why not? What if only five plants survive? What if 100 plants survive, but there are no new seedlings being produced?
3 If the goal is to create a population of 50, 500, or even 5 000 reproducing plants, how many seeds or plants should be used in the experiment?
4 On the basis of your experience with creation of new populations of plants, discuss the relative values of reintroduction versus protecting the remaining populations

of declining species. What logistical, political, or biological considerations should be taken into account?

5 On the basis of the data from your projects so far, what new information or techniques should be incorporated in future projects of this kind?

6 If you have worked primarily with the reintroduction of locally extirpated species that are not of conservation concern (i.e. not threatened or endangered), what additional considerations would you have to take into account if you use the same techniques with globally threatened species?

7 What have you learned from the plant experiments that is applicable to animal reintroductions? What would be different about an animal reintroduction?

Making It Happen

Experience gained in these initial experiments can be more widely applied. As a class project, you might consider restoring an entire biological community and documenting the process. The site could be abandoned farmland, a lawn, a polluted pond, a channelized stream, a tree plantation, a vacant lot, or overgrazed rangeland. Such a project will probably take many years, but would serve as a demonstration to a wider audience. Volunteer workers from school clubs, local conservation organizations, and the general public could provide needed labor in creating fences, preparing the site, putting in new plants, and long-term maintenance. A good photographic record is useful in documenting the project and in writing popular articles about the project. Another project could be to focus on one species that is rare or endangered in your region. Discussions with officials in the Heritage Programs, or the regional office of state conservation agencies might help to identify suitable species. A comprehensive project would involve study of the ecology of this species in its existing populations, developing methods for propagating it in gardens or greenhouses and finally establishing new experimental populations in the wild.

In all such projects, make sure to seek the advice of local botanists or naturalists, who will have a good knowledge of local habitats and species, the locations of possible source populations (for seeds or transplant material), and may have valuable knowledge of recent changes in land use or the population dynamics of particular species.

Further Resources

The key reference in this field is Falk et. al. (1996). Guerrant et al. (2004) is also an important supporting resource. Useful collections of articles can be found in the journals *Restoration Ecology* and *Restoration and Management Notes*.

CHAPTER 18

Edge Effects: Designing a Nest Predation Experiment

Malcolm L. Hunter, Jr.

> Game is a phenomenon of edges. It occurs where the types of food and cover which it needs come together, i.e. where their edges meet. Every grouse hunter knows this when he selects the edge of a woods, with its grape tangles, haw-bushes, and little grassy bays, as the likely place to look for birds. The quail hunter follows the common edge between the brushy draw and the weedy corn, the snipe hunter the edge between the marsh and the pasture, the deer hunter the edge between the oaks of the south slope and the pine thicket of the north slope, the rabbit hunter the grassy edge of the thicket.

The person who wrote those words, Aldo Leopold, is considered by many to be one of the true pioneers of conservation biology. He was also an avid hunter, and when he wrote those words in 1933 he was speaking as a human predator. He knew that edges often attracted the prey that he sought. Non-human predators seem to be very familiar with this phenomenon as well and, while this might be a good thing for foxes and crows, it may be a decidedly bad thing for the animals that are prey. In particular, conservationists have expressed concern that as landscapes are fragmented more and more, many song birds end up nesting in edge environments where they are subject to unnaturally high levels of predation.

To understand this phenomenon many researchers have studied patterns of predation near edges using field experiments (Batary & Baldi 2004) and this exercise provides you the opportunity to undertake such an experiment yourself.

Objective

- To design and execute an experiment to estimate the risk of bird nest predation at different distances from an edge.

Procedures

This exercise is based on the idea that if you place many baits at different distances from an edge and monitor the portions of them that are consumed, these portions will constitute an index of the predation pressure from potential bird nest predators. For example, if 60% of the baits placed within 2 m of an edge disappear compared to 15% for those that are 75 m from an edge, this would indicate that nests 2 m from an edge are four times more likely to be preyed upon than nests 75 m from an edge.

Design

For this exercise we will be taking a design-your-own-experiment approach. The first issue you need to consider is what type of bait to use. Obviously eggs are a reasonable facsimile if you are interested in nest predation, although the most readily available ones – chicken eggs – are rather conspicuous and may be too large for some potential predators (e.g. squirrels) to open. Consequently, most researchers have used quail eggs for these types of experiments because they are much smaller and covered with brown blotches that make them less conspicuous. Some researchers believe that even quail eggs are too big to be available to all potential predators of song bird nests (e.g. mice) and thus they have used eggs from domestic finches. In short, this experiment will probably work best with the smallest eggs that are available to you. However, any broadly attractive bait – even chocolate eggs or peanuts – would probably work.

The next issue is where to do the study. Most predation-edge research has been undertaken in forests that are bounded by fields, clearcuts, power-line rights-of-way, and similar environments. To make your experiment less redundant with previous research you could work in a forest adjacent to a river, lake, marsh, or road, or place your baits in both a forest and early-successional ecosystem (i.e. have transects on both sides of the edge). It would increase the statistical rigor of your work if you used many different sites with the same basic edge type, but logistical constraints may confine you to one site.

Next, you need to decide how to distribute the baits at each site. Statistically, it would be best to distribute the baits randomly and, after the fact, measure their distance to the nearest edge. However, it is much easier to place baits along transects, either parallel or perpendicular to the edge. The optimum configuration will depend on whether you are working with one large site or several small sites, whether or not the edge you are studying is relatively straight, the size of your class, and other considerations. The bottom line is that you must know the distance from each bait to the edge; you should have at least 50 baits that are relatively close to the edge and 50 that are relatively far. Be sure that the farthest points are not actually closer to a different edge. The distances between adjacent transects or between bait points along a transect should not be too small (e.g. < 15 m). However, there is an unavoidable trade-off here because if you use a short interval you can set out more baits, thereby increasing your overall sample size. On the other hand, if your baits are farther apart your results are less likely to be biased by sampling from a small, relatively homogeneous environment (e.g., dominated by one vegetation type or heavily influenced by the behavior of a single animal that happens to like your baits and learns how to find them).

Execution

When you set out the baits, each point along your transects needs to be marked with something that will definitely allow you to relocate it, but without attracting attention from potential predators or vandals. Baits should be located near, but not at, the transect point using some simple rules that will randomize the direction from the point. Choose a spot that is somewhat concealed (e.g. in a patch of ferns), but not too concealed (e.g. inside a hollow log). It is very important that you write a very detailed description of where the bait is so that if it is gone when you return you will be certain that you are looking in the right spot. Putting something small and inedible under each bait, such as a small coin, would confirm your relocations.

It is often useful to know if the density of vegetation around a nest might affect the likelihood of it being preyed upon. Try to devise a simple, consistent way to estimate the vegetation density near your baits. You might want to do this at two different scales; e.g. within 1 m and 5 m of the bait.

After the baits are in place you need to wait a reasonable period before you check them. If you check them too soon, very few will be gone and it is hard to analyze a data set full of zeros. However, if you wait too long almost all of them will be gone and it is equally hard to analyze a set of 100 percents. Most researchers run these experiments for 2 weeks with a check at one week to make sure that the baits are not disappearing too fast. When you check the baits you will find some completely gone and some untouched and some in an in-between state (e.g. moved a short distance, showing teeth marks, pecked open). Record the evidence in detail and bring any remaining fragments back to the lab; one can often distinguish avian and mammalian predators from beak marks versus teeth marks and some times teeth marks will reveal the particular species.

Analysis

Back in the lab you will need to pool all the data from the different teams for statistical analyses. If your sample size is large you may be able to pool your data into three or more categories (e.g. baits < 15 m from an edge, 15–60 m, 60–120 m, and < 120 m) rather than just two (near versus far). Similarly you may also be able to do separate analyses for nests that probably were preyed upon by birds versus those preyed upon by mammals, or those that were above or below some threshold in terms of the density of the surrounding vegetation.

The simplest way to make a statistical test of an "edge effect" on predation rates is by analyzing your data in a contingency table. A contingency table is a tabulation of data from one or more samples that permits you to test for differences in frequencies between statistical samples. In this experiment you are comparing differences in the frequency of predation between samples from two habitat treatments, i.e. edge habitats and interior habitats.

You want to test the null hypothesis that the frequency of predation does *not* differ between habitat treatments. Contingency table analysis will test this null hypothesis by estimating the probability that the frequency of predation does not vary by habitat treatment. If the probability is very small (say, less than 0.05) then this hypothesis is likely not true, and you can safely conclude that predation does not vary by habitat treatment. If the probability is high (generally, 0.05–1.0), then you will not reject the null hypothesis and can conclude that there is no edge effect.

The first step in making this test is to tally your data into a table with two rows and two columns. Denote the two rows as "not depredated" and "depredated" and denote the columns as "edge" and "interior." Now tally your data in the appropriate cells in the table. For example, if you put 100 baits out, and 30 or 50 baits placed in edge habitats were eaten, then put 30 in the table cell corresponding to "edge-depredated" (see Table 18.1). This is the observed frequency, *f*, that baits occurred

Table 18.1 Observed frequencies.

	Edge	Interior	Total
Not depredated	30	41	71
Depredated	20	9	29
Total	50	50	100

in this cell of the contingency table. If you subdivided your analysis into different distance classes from edges, you can still do this analysis by adding extra columns.

If the null hypothesis were true (frequency of predation is comparable in different habitats) then the frequency of baits expected, F, in each of the cells in the contingency table would be:

$$F = RC/n$$

where R and C are the row and column totals, respectively, associated with a given cell, and n is the total number of baits put out in the field. Construct a similar table with the expected frequencies. Table 18.2 presents expected frequencies corresponding to the observed frequencies on Table 18.1 if predation frequency was independent of habitat type:

Now you can calculate a statistic, the chi-square statistic (χ^2), that will measure how far the observed frequencies in Table 18.1 depart from the expected frequencies in Table 18.2. If the total departure is large, then obviously the observed frequencies differ from the expected ones, and your null hypothesis is probably wrong. If the total departure is small, then the null hypothesis probably fits your data nicely. To calculate this departure, calculate across the rows and columns:

$$\chi^2 = \Sigma(f - F)^2/F$$

For these data, the value for the first cell is $(30-35.5)^2/35.5 = 0.85$ and χ^2 overall $= 0.85 + 0.85 + 2.09 + 2.09 = 5.88$. (Technically for small 2×2 tables such as this one, you should add a small correction factor, Yates correction for continuity, for each cell, such that $\chi^2 = \Sigma(f - F - 0.5)^2/F$; but this will be unnecessary for contingency tables larger than 2 rows by 2 columns).

Last, you need to calculate your degrees of freedom (df) that reflect how your data are organized in the contingency table:

$$df = (\text{no. rows} - 1)\ (\text{no. columns} - 1)$$

In this case $df = 1$. The critical value of χ^2 that corresponds to this number of degrees of freedom can be found in the first row of Table 18.3. If the calculated value of the χ^2 statistic is larger than the critical value in Table 18.3, then you can say that there is a less than 1 in 20 chance (i.e., 0.05) that the null hypothesis is true. This is a small

Table 18.2 Expected frequencies corresponding to observed frequencies in Table 18.1.

	Edge	Interior	Total
Not depredated	35.5	35.5	71
Depredated	14.5	14.5	29
Total	50	50	100

Table 18.3 Critical values of the Chi-square statistic at $\alpha = 0.05$.

DF	1	2	3	4	5
χ^2	3.8	6.0	7.8	9.5	11.1

probability, so if the χ^2 is larger than this value then you can safely say that predation frequencies varied among habitats and hence there is an "edge effect." This is clearly the case for the data in Table 20.1, for which the χ^2 was 5.9, which is larger than the critical value for 1 degree of freedom of 3.8. Critical χ^2 values for different degrees of freedom have been included in Table 20.3 in case you subdivided your analyses by various distances from the edge. Any general statistics book will have a more complete table of χ^2 if you want to explore the significance of your data at other probability levels.

Expected Products

- This may be as simple as a table of your data and an analysis of them or an entire research paper prepared for a particular journal.
- Responses in a form indicated by your instructor to the Discussion questions below.

Discussion

1 Clearly there are some significant differences between baits and real bird nests. What are they and how would they affect your results? Which ones could be minimized if you were to repeat the experiment with fewer logistical constraints?
2 Discuss the overall landscape in which you conducted your experiment. How might your results have been affected by land use patterns within a 5-km radius of your study site? What do landscape patterns in your region as a whole suggest about the overall impact of edge effects?
3 Given the spatial patterns of the sites you studied, their type of management, and the objectives of their owners, how could you reduce a predation-edge effect on these sites? Are these ideas widely applicable in your region?

Making It Happen

This project could present an opportunity to generate a useful scientific paper for publication in a journal. Because many such experiments have now appeared in the literature, along with some criticism about their naturalness, you need to have a rigorous experimental design (with hundreds of baits distributed over many sites) and to tackle a new angle on the topic. Repeating an old experiment in a new state or country is probably not new enough; birds and their predators do not recognize political boundaries. Review the existing literature and see if you can find a new type of edge to explore. It will be difficult to gather a robust data set in one field season but perhaps your instructor will be willing to coordinate a multi-year effort.

Even if your results do not merit publication in a major journal, they may well be of interest to the people who manage the land that you studied, especially if you found high levels of predation and have some suggestions of how to mitigate them.

Further Resources

Many nest predation experiments have been reported in the literature and some review papers (e.g. Hartley & Hunter 1998, Stephens et al. 2003, and Batary & Baldi 2004) will give you a good entree into the literature. For further insights and critiques see articles in *Conservation Biology* 18(2) on the techniques used in bird nest predation studies. Lindenmayer and Fischer (2006) provide a comprehensive treatment of the fragmentation issue.

PART 5

Ecosystems and Landscapes

CHAPTER 19

Ecosystem Fragmentation: Patterns and Consequences for Biodiversity

James P. Gibbs

Take a tour of the landscape you inhabit in a small plane, up at about a thousand meters, and you are not likely to see unbroken forests or grasslands stretching off to the horizon. Most of the places where people live are a patchy mosaic of woodlots, fields, shopping malls, college campuses, and so on. Unfortunately for most of the species that have to share these landscapes with us, our activities have often reduced natural ecosystems to small, isolated islands in a sea of human-altered lands. To put it more formally, many species are confined to patches of suitable habitat that are isolated in a matrix of ecosystems that are not habitat. Habitat loss and fragmentation together is clearly the biggest threat to biodiversity. This said, the habitat loss and fragmentation process is a complex one and predicting its effects is not at all straightforward. The fundamental question we address here is: Can landscapes be fragmented in such a way that permits humans and biological diversity to coexist?

Objectives

- To explore through a mapping exercise what happens to a neotropical forested landscape as it undergoes the fragmentation process
- To predict what will happen to the biota residing within the landscape as a result of these changes.

Procedures

Overview

The first part of the exercise involves measuring changes in a neotropical forested landscape during the process of fragmentation. You begin with a blank grid that represents an undisturbed landscape dominated by forest. Much of the forest is on the upland but some also occurs in wetlands that are connected by streams which are themselves surrounded by gallery forest. Both wetland forest and gallery forest are considered "seasonally inundated forest" and are both indicated on the map by wetland symbols. Starting with the blank grid, you will mark grid squares in a progression that mimics fragmentation of the landscape associated with colonization of it by humans, first by adding a major road and the cleared lands associated with it, and then adding secondary roads and tertiary roads and the cleared lands associated with them. Filled grid squares will represent areas cleared of forest and converted to agricultural purposes whereas cross-hatched grid squares will represent edge zones of remaining forest that is directly adjacent to cleared areas. You will then repeat this fragmentation process while invoking some simple land use guidelines to examine how they might influence the outcome in terms of structure of the landscape and the biodiversity within it. You will end up with 3 landscapes to compare: (i) the original landscape, (ii) the landscape subjected to uncontrolled fragmentation, and (iii) the landscape subject to fragmentation guided by some simple land use regulations and alternatives.

The second part of the exercise enables you to predict what will happen to the biota residing within the landscape as a result of fragmentation. For each of your 3 mapped scenarios, you will calculate some key biological parameters to make predictions about the state of biological diversity and ecosystem functions within the landscape. You will examine how changes in the landscape affect (i) ecosystem diversity, (ii) species diversity, (iii) ecosystem function in terms of carbon sequestration, (iv) population viability of a large herding mammal, (v) foraging energetics of wide-ranging birds, and (vi) genetic drift in a canopy tree. By contrasting these biological indicators in the 3 landscapes you generate, you will get a good sense of how fragmentation affects biodiversity and how we can mitigate some of its negative effects through planning and incentives.

Getting Familiar with the Map

First, get oriented to the basic map (Figure 19.1) by noting the cardinal directions. Which way is north? South, east and west? Second, familiarize yourself with the map's scale. Each grid square is 100 m on a side. What is the area of each grid square? What is the width and length of the study landscape? What is the total area of the landscape (in ha and km^2)? If you move horizontally from the center of one grid square to the center of the next, what distance have you moved? If you moved diagonally, how far have you moved? Now look at the different cover types. Can you recognize the inundated (wetland and gallery) forests? The upland forests? The streams and other watercourses?

Carrying Out Scenario 1: Original Landscape with Natural Small, Scattered Disturbances and a Low Human Population

Scattered disturbances are typical even within the original, unfragmented landscape. These might be due to lightning strikes that have created small openings. They also may be small, shifting garden plots or areas where humans have lit fires to generate secondary growth and attract game animals. These disturbance patches usually don't comprise a large portion of the landscape (perhaps up to 2% of the area) and are generally well dispersed. To mimic this situation, on Figure 19.1 randomly choose 2% of the grid squares and convert them to early successional vegetation (filled

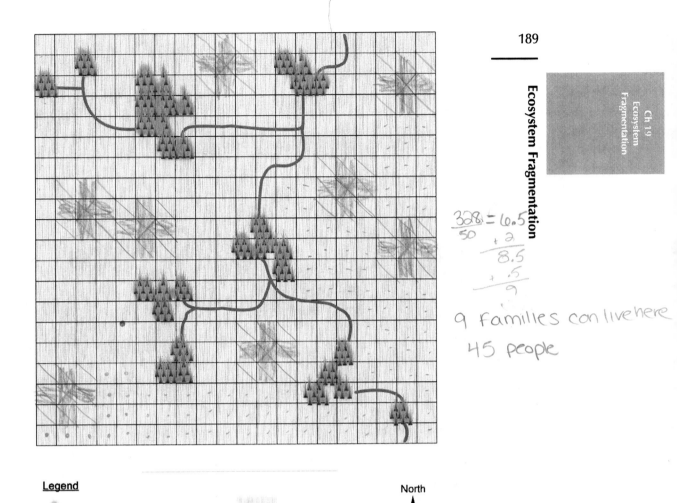

$$\frac{328}{50} = 6.5$$

$$+ 2$$
$$\overline{8.5}$$
$$+ .5$$
$$\overline{9}$$

9 families can live here

45 people

Legend

= Seasonally inundated forest   = Upland forest

) = Stream

1 hestacre

☐ = 10,000 m²

North ↑

Fig. 19.1 Base map for ecosystem fragmentation scenario 1.

grid squares) and change the grid squares surrounding these to edge environments (cross-hatched grid squares).

What is the human population being supported within the landscape? Assume that each family (average of 5 people per family) needs exclusive access to 3 hectares of cleared land for cultivation to meet their needs or 50 hectares of forestland (upland or inundated and not necessarily contiguous) for extraction of natural products and hunting. Often people combine both cultivation and harvest of wild products but we will consider a simple division of livelihoods in this case.

Carrying Out Scenario 2: The Landscape Fragmented in an Uncontrolled Manner

Starting with a blank map of the original landscape (Figure 19.2), add a road dissecting the region. This road might be the end result of exploration of a remote

187 available land
= 3.5

3 families +
54

57 families =

285 people

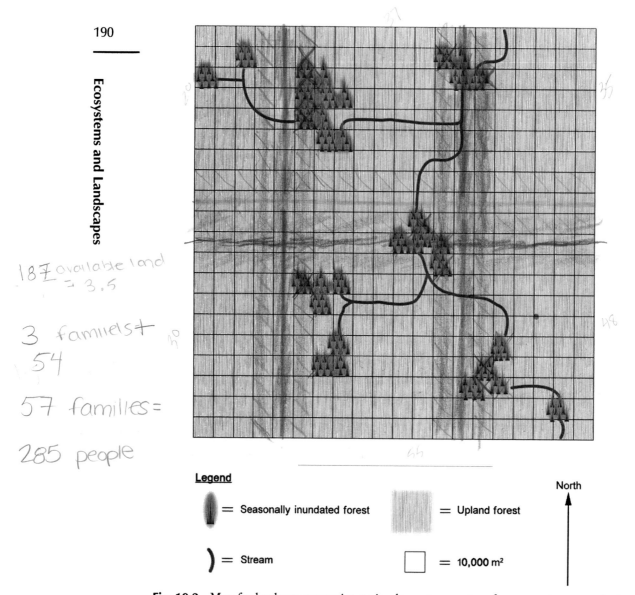

Legend

= Seasonally inundated forest

= Upland forest

) = Stream

☐ = 10,000 m²

North

Fig. 19.2 Map for land use conversion to implement ecosystem fragmentation scenario 2.

area for oil or gas or the result of a government effort to access a frontier zone. To make this road, simply draw a heavy line along an east-west axis through the middle of one of the central rows of grid squares. The road width is negligible and can be ignored but it provides immediate access to the grid squares traversed and those immediately adjacent to them (150 m back from the road). These are converted to agriculture. Therefore, fill in all traversed and adjacent grid squares - three rows total.

Note that this initial dissection of the forest also changes the forest that is adjacent to the converted lands into edge habitat. Consider that ecologically important edge effects can extend at least 100 m into a stand, so cross-hatch all the forested grid squares adjacent to cleared land to indicate where forest edge has been created.

Now add more roads. These are the kinds of roads created by people who follow the first major road and now seek to colonize the area and convert more of the

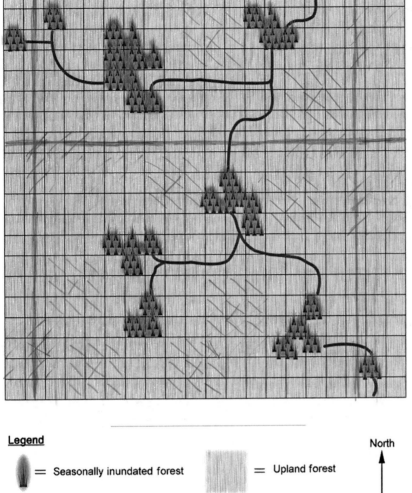

8 families

58 families

Legend

= Seasonally inundated forest   = Upland forest

) = Stream   ☐ = 10,000 m²

North ↑

Fig. 19.3 Map to implement ecosystem fragmentation scenario 3.

forest for agriculture and other uses. Add two such roads by drawing dark lines perpendicular to the initial road that cross it at about 450 m and 1450 m along its length. Extend the secondary roads in both directions right across the landscape. Repeat the process of demarking converted lands 150 m from the roadsides and then the edge habitats adjacent to them.

Calculate the human population supported within the landscape.

Carrying Out Scenario 3: The Landscape Fragmented Following Some Simple Land Use Guidelines

Starting with a clean map (Figure 19.3), add roads and cleared lands as you did in Scenario II, with the objective of generating sufficient resources for a comparable number of people, but place the roads in any configuration you want that satisfies these simple land use guidelines and land use alternatives (adapted in part from Laurance and Gascon 1997):

- Triple the production rates on cultivated land through provision of fertilizer and perhaps alternative crops so that local people now need to clear only a third as much forest to meet their needs. In other words, each family can meet its needs on just one hectare of cultivated land because productivity has been tripled. Thus you only need to convert the lands up to 50 m from roadsides, so fill in only the blocks directly intersected by the road.
- Prohibit forest clearing of any habitat block that contains a watercourse. Note that roads can traverse watercourses but the forest blocks that include watercourses are not cleared.
- Protect all rare ecosystems – no conversion of inundated forests.
- Allocate half of the landscape to production purposes and human use while allocating the other half to reserve status.

Recalculate the human population supported within the landscape.

Landscape Analysis of Ecosystem Diversity

For each of the maps that you produced for the 3 fragmentation scenarios, tally the area of the landscape that is upland forest interior, upland forest edge, inundated forest interior, inundated forest edge, and land converted from forest to agriculture. Note that riparian zones of the watercourses are natural "edges" or ecotones, but we are concerned here with forest edges adjacent to agriculture and burnt patches. Calculate the fraction of the landscape composed of each ecosystem type. Last, estimate ecosystem diversity within each of these landscapes using the Shannon-Weiner index of diversity, which is

$$-\Sigma p_i \times \log(p_i), \qquad H' = -\Sigma p_i \ln p_i$$

where p_i = the fraction of the landscape represented by ecosystem$_i$. For example, if interior forest occupied 900 of the total 1 000 grid squares and inundated forest occupied the rest of the landscape, then Simpson's index of diversity would be

$$-1 \times (0.9 \times \log(0.9)) + (0.1 \times \log(0.1)) = 0.325.$$

Your calculations will be similar but made across all ecosystem types.

Estimating Changes in Ecosystem Function: Landscape-Level Carbon Sequestration

1 ha = 150 tons of carbon

Based on the estimates of Laurance et al. (1997) for Amazonian moist forest near Manaus, Brazil, forest biomass averages 300 tons/ha with carbon comprising 50% of that amount. Forest clearing commits 95% of forest biomass to carbon emissions from burning and decay with 5% remaining as relict living trees in pastures or inert as charcoal in the soil. Forest edges lose 10% of their biomass because mortality of trees is higher on the edges. Do not distinguish inundated forests from upland forests for this exercise. Based on these relationships, estimate the tons of carbon sequestered by the landscape in above-ground woody biomass under the 3 different landscape scenarios.

Estimating Changes in Faunal Diversity

Estimate the faunal diversity of an average forest block within the landscape under the different scenarios of fragmentation (undisturbed, uncontrolled fragmentation, managed fragmentation). This can be done with information on birds, mammals, frogs, and ants gathered by Gascon et al. (1999). These researchers worked over two decades to determine which species near Manaus, Brazil, primarily used forest interior, forest edge, and "matrix" or open lands near fragments. Note that some of the forest species listed below use the matrix but still rely on the primary forests.

Approximating from the Gascon et al. (1999) report (Figures 19.2 and 19.3), we will assume that our study area harbors:

- 123 bird species, 31 of which use the matrix (converted lands) versus 92 that use the forest edge and 92 that use forest interior. Note that no birds were restricted to the forest interior, so the same 92 occur in both edge and interior areas.
- 62 frog species, 16 of which occur only the matrix, 52 of which occur in the forest interior, and 51 of which use the forest edge
- 15 mammal species, 4 use the matrix only, 15 use the forest interior, and 10 use the forest edge
- 127 ant species, 32 use the matrix only, 104 use the primary forest, and 44 use the forest edge.

Now calculate the average diversity per hectare for each of these faunal groups for each of the 3 scenarios. Do not distinguish inundated from upland forests for this part of the exercise. To do so, for each landscape multiply the number of hectares of each ecosystem type by the expected diversity within it. Next sum these values across types. Last, divide by the sum of the weights, which is the same as the total area of the landscape. For example, if the landscape was composed of 1 000 grid squares, of which 500 were primary forest, 250 were forest edge, and 250 were matrix, then ant diversity on average in that landscape would equal

$$((500 \times 104) + (250 \times 44) + (250 \times 32))/1000$$

This **weighted average** will indicate how many species are likely to occur, on average, per hectare in each landscape.

A Simple Population Viability of a Herding Species with a Large Home-Range

Assume that a tract of forest (interior and edge) of at least one km^2 is required to support a herd of white-lipped peccaries and that these animals live at densities of five peccaries per km^2 (Fragoso 1998). Also assume that the peccaries are reluctant to cross roads and cleared areas because they will be shot and therefore they restrict their movements to individual forest remnants. It follows that isolated forest blocks support isolated populations.

What is the total peccary population in each scenario? To answer this, you will first have to tackle the following questions: How many peccaries can each of the remaining forest patches support? What fraction of patches contains both the upland and wetland forest that are required to meet the annual needs of these animals during the wet and dry seasons? Note that if at least one hectare of both wetland and upland forest are not available within a forest patch, then a population cannot be supported.

Assessment of Mobile Species and Foraging Energetics

Let's consider a wide-ranging, large-bodied frugivorous bird that must visit many sites every day to harvest newly ripened fruit. A toucan, quetzal, or large parrot are good examples. Let's assume the fruit tree species occurs only in inundated forests, that is, in the highly clumped distributions that are typical of many tropical trees. Assume that a pair of these birds must visit 5 such patches (a "patch" equals one hectare block of inundated forest) each day. How far must these birds travel on average each day to meet their daily needs in each scenario?

To answer this question, first trace the shortest path possible between all patches of inundated forest in the landscape. Start with an isolated patch in one of the corners of the landscape. It's virtually impossible to find the exact shortest path so just try to link all patches together as might a foraging bird that was trying to save energy flying between all the patches. As you move to the next nearest patch of inundated forest not yet visited, sum up the distance of each sequential move. The total distance traveled divided by the total number of patches visited equals the average cost of accessing a foraging site. Recall that the birds must visit 5 patches per day to meet their needs.

Changes in Genetic Diversity of a Canopy Tree

Consider a rare tree species with mature individuals distributed evenly through the upland forest at a density of only one per hectare. A good example is *Pithecellobium elegans* (Chase et al. 1996, Hall et al. 1994). Any trees within 250 m of any other trees represent part of the same breeding population (i.e. they can participate in gene flow). More distant individuals are unable to exchange pollen effectively and are therefore considered to be members of a different population. First, mark the individuals linked through potential gene flow by drawing a line that includes all collections of individuals within 250 m of another individual.

Next, assuming that the genetically effective population size in this species equals the census population size (only breeding adults in this long-lived species are considered), what average fraction of heterozygosity will be lost over the next 100 generations from each of the remaining breeding populations? Recall that the formula for estimating the amount of genetic variation (heterozygosity) in a population of size N_e retained after t generations equals

$$[1 - (1/(2 \times N_e))]^t$$

Repeat these calculations for each *Pithecellobium elegans* population under each of the 3 landscape fragmentation scenarios. What is the least amount of genetic diversity lost by any single population under each scenario?

Expected Products

- A table that summarizes, for each of the fragmentation scenarios, (i) the human population supported in the landscape, (ii) the characteristics of the landscape (fraction of land in different ecosystem types), (iii) estimates for these key biological indicators:

- fraction of land in different ecosystem types
- ecosystem diversity at the landscape level
- carbon emissions from the landscape
- average faunal diversity per block
- population size for white-lipped peccaries
- foraging energetics for the frugivorous bird
- genetic diversity in the tree *Pithecellobium elegans*.

- To indicate how sensitive each parameter is to the fragmentation process, a calculation of the proportional change in each parameter relative to its value in the original landscape (Scenario 1)
- Responses in a form indicated by your instructor to the Discussion questions below.

Discussion

1 Can you conclude that biodiversity and substantial human populations can co-exist despite loss and fragmentation of native habitats in the landscape?
2 What social incentives and political means are available to actually achieve a landscape like that in scenario 3?

Further Resources

We recommend Lindenmayer and Fischer (2006) for an overview of ecosystem fragmentation issues and the various citations herein for further details.

Due Tuesday:
Hard Copy
1) Maps
2) Table
3) Typed responses to questions below table

4) Read online textbook ch. 4 + 5

CHAPTER 20

Forest Harvesting: Balancing Timber Production and Parrot Habitat

Malcolm L. Hunter, Jr., and Robert Seymour

"Payroll or pickerel" may be a pithy little expression, but it grossly simplifies the complex choices that usually face natural resource managers when they try to manage an ecosystem to produce a commodity while still maintaining biodiversity and other ecological values. This dilemma is especially conspicuous in forests for two reasons. First, there is a lot at stake in forests: thousands of species can occupy a hectare of forest ... as can thousands of dollars worth of timber. Second, the impacts of cutting timber are often highly visible because forests tend to show their wounds, whether inflicted by a natural fire or a chainsaw. Think about any timber-cutting operation you have seen – even a rather light one – and compare it to looking at an overfished lake or an overgrazed grassland.

Much has been written about how to maintain biodiversity in forests that are being actively managed for timber production (e.g. Hunter 1999, Lindenmayer & Franklin 2002). Sometimes focused decisions are required such as which species and sizes of trees to cut using what kind of equipment. These are questions that can be evaluated in the context of how they will disturb the forest and how such artificial disturbances will differ from the effects of natural disturbanes such as fire and windthrow. Sometimes, forest managers stand back and ask broader questions such as how much of the landscape to allocate to intensive timber production, how much should be reserved for biodiversity and other ecological values, and how much should be managed for multiple, competing objectives. Caring about both timber production and biodiversity means walking a high wire in a delicate balancing act.

Objective

- To undertake some basic analyses behind allocating land to different types of forest management as a vehicle for understanding the key issues involved in managing ecosystems to balance production of a commodity while maintaining biodiversity.

Procedures

This exercise will involve dividing into teams to tackle a forest management problem and then coming back together to compare your solutions. The problem is complicated enough to generate a number of different solutions, although actually it is greatly simplified compared to a similar problem in the real world.

You will play the role of a team of forest managers employed by a small timber company that owns a sawmill and a 30 000 ha forest. The company owners have signed an agreement with the government to help provide habitat for an uncommon species of parrot. This is part of a cooperative effort to keep the parrot population from declining to a point where it will be necessary to list it as an endangered species, at which point many restrictions would be imposed on your company's operations. The agreement calls for your company to provide enough habitat to support a population of 100 breeding pairs of the parrot. Your supervisor has asked you to develop a plan that will meet this goal while producing at least 120 000 cubic meters of wood per year to supply the saw mill.

First, let's consider the parameters or constraints within which you must work, starting with the parrots. Each parrot pair requires 80 ha of native forest that is over 150 years old in which to breed. Outside of the breeding season each pair requires a total of 120 ha of native forest, but it only has to be over 75 years old. To minimize the chance of a catastrophe wiping out the whole population, it is preferable to have the parrots well-distributed across the forest. This will require good habitat connectivity to allow juveniles to disperse throughout the forest. The key to maintaining habitat connectivity is to focus on the distribution of roads and openings where parrots are at some risk of being shot by poachers.

Second, let's turn to the trees: you are starting with 30 000 ha of native forest, all of which is over 150 years old and thus suitable parrot habitat. (You only need to determine how this forest will look after your management plan is fully implemented. Figuring out the transition from its current state to its future state is important, but beyond the scope of this exercise.) You are in a position to grow, process, and market just two tree species, one native and one exotic. The native species is a kind of mahogany which is usually grown in small, strip clearcuts (50 m × 200 m) on a 100-year rotation. In other words, a strip is cut and then allowed to grow for 100 years before being cut again. When grown on a 100-year rotation a mahogany harvest yields 500 m^3/ha; in other words, over a 100-year rotation the average growth of mahogany is 5 m^3/ha/year. Occasionally, mahogany is grown on a longer rotation of 200 years. Under this management, growth averages a bit less, 4 m^3/ha/year, and thus the yield after 200 years is 800 m^3/ha (4 m^3/ha/year × 200 years).

The exotic tree is a species of eucalyptus that is grown in plantations on a 25-year rotation (i.e. stands of at least 5 ha are planted then clearcut 25 years later). Despite the short rotation, harvests of eucalypts yield 500 m^3/ha because the eucalypts grow very fast. Over a 25-year rotation they average 20 m^3/ha/year.

The forest is square in shape (about 17.3 km × 17.3 km) and the mill is located along the middle of the south side. You will need a permanent road system to extract the wood. Road construction and maintenance costs are substantial (both economically and ecologically) but the financial costs can be covered by your timber revenues.

How do you tackle this problem? To keep things simple, let's think of this as primarily an exercise in allocating different parts of the forest to one of four different types of management:

- mahogany grown on a 100-year rotation
- mahogany grown on a 200-year rotation
- eucalyptus plantations
- leaving the native forest as an uncut ecological reserve.

The first question to ask is, would allocating all of the forest to one of these management types meet the goals? Obviously having 100% eucalypt plantation or 100% reserve is not an option because they provide no parrot habitat and no timber harvest, respectively. The 100-year mahogany management would only provide parrot habitat outside of the breeding season. The 200-year mahogany management would both generate timber and provide some breeding habitat, but is it enough of each? For your first calculations, estimate the total volume of timber that would be produced each year under a 200-year rotation of mahogany and the number of parrots that can be supported under this management regime. If after 10 minutes you are stuck, refer to the textbox entitled **Help!** at the end of the exercise where we have provided some further guidance.

Once you have figured out the basics of calculating timber productivity and the availability of parrot habitat, the only thing holding you back are the constraints listed above and your creativity. (But don't let your creativity trample the bounds of what is likely to be realistic and financially feasible: e.g. no helicopter logging). Try several different allocation strategies and discuss the pros and cons of each within your group. Once you have settled on your preferred choice draw a simple map showing how the different management types would be distributed on the forest and how you would access these areas with roads. Finally, present your plan to the rest of the class.

Expected Products

- A table showing how much land you will allocate to each management alternative and much timber will be produced annually and how many parrots supported (include estimates for both breeding and non-breeding habitat).
- A map showing how the different land uses are distributed and the road network.
- Responses in a form indicated by your instructor to the Discussion questions below.

Discussion

1 The forest described here is far too perfect to be real. Among other things, unpredictable events will always intervene to unbalance the age-class structure of a managed forest. One year a fire bums 5 000 ha. Another year, a fire burns the mill and no wood is cut for the next 2 years until the mill is rebuilt. How

would you build a "safety cushion" into your plan that would protect both the parrots and your timber productivity from catastrophes?

2 The landscape context of a forest is always a consideration for ecologically sensitive forest managers. What did you assume about land use in the area surrounding your forest? How would your management differ if your forest were embedded in a landscape of similar forests versus an agricultural landscape?

3 If you could grow and harvest mahogany using individual-selection harvesting (where you found and cut each individual tree as it reached maturity and thus created only small openings in the forest canopy), what would be some of the likely advantages and disadvantages of this approach?

4 How should society assure that an appropriate balance is struck when faced with choices like the ones outlined here?

Making It Happen

Forest managers everywhere are faced with trade-offs like the ones presented here; with a little bit of digging around, you are likely to find a local issue that would merit some analysis. If the forest is publicly owned there is likely to be an explicit mechanism for offering your input. If it is privately owned, you will have to be extremely discreet and tactful in approaching the owner, or risk having an unpleasant confrontation on your hands. In any case, be prepared to do a lot of legwork because the issues are far more complex than in this simplified example, and a superficial analysis can easily do more harm than good.

Further Resources

For a broad treatment of balancing timber production and biodiversity, see Hunter (1990, 1999) and Lindenmayer and Franklin (2002). To get a more realistic look at silvicultural options and how foresters balance the age-class structure of forests, read Smith et al. (1997), especially pages 425–431.

See overleaf for help with the exercise on p. 198.

Help!

Estimating the annual timber productivity of any particular management strategy is simple. All you have to do is multiply the number of hectares allocated to each type of management by its annual timber productivity; in this first case 30 000 ha times the annual productivity of 200-year rotations of mahogany. Determining how much parrot habitat will be provided by a particular regime is a bit more complicated. Again, we can assume that all the reserves are breeding habitat and that none of the plantations is. Thus the key is how mahogany management affects the age-class structure of the forest. For example, if you are managing the entire forest on a 200-year rotation and wish to have a stable harvest each year, then each year you will cut 1/200th of the forest. This means that at any given time 1/200th of the forest will be 1 year old, 1/200th will be 2 years old, and so on up to 1/200th being 200 years old (until the day it is cut and becomes 0 years old). We can lump these age classes together into more convenient categories; for example, at any given time, 5/200ths of the forest will be from 1 to 5 years old. Alternatively, we could say that half the forest (100/200ths) will be from 1 to 100 years old and half will be from 101 to 200 years old. Using this perspective you should be able to figure out how much of a forest managed on a 200-year rotation will be over 150 years old, and thus suitable parrot breeding habitat, at any given time. It may help to graph this with histograms along the x-axis representing each age class, and the area of each age class along the y-axis. Once you know the amount of breeding habitat you can divide by the breeding range size of each parrot pair to estimate the number of parrots. Do not forget to make sure there is enough nonbreeding habitat too.

CHAPTER 21

Protected Areas: A Systematic Conservation Planning Approach for Ecoregions

Pablo I. Ramírez de Arellano

In the tool box of biodiversity conservation, establishing and managing protected areas is one of the dominant strategies and this requires deciding which areas are most worthy of conservation action. Ideally, these areas should include the known biodiversity of the region in which they are situated, and they should protect biodiversity from processes that threaten its persistence (Margules & Pressey 2000). But the total area of land that can be set aside, or otherwise managed, for conservation is often severely limited by social and economic constraints. Care must therefore be taken to direct scarce resources to areas of highest conservation priority, defined in terms of both conservation value and degree of threat (Margules & Pressey 2000). Since the 1980s, several algorithms and techniques have been developed to select new reserves that return the greatest biodiversity benefit (species protected) for the funding invested in securing habitat. One prudent approach is to choose a set of sites that achieves comprehensive representation of biodiversity for the minimum cost (Possingham et al. 2000). This is called the "minimum representation" problem. If the cost of protecting land is a linear function of the number of sites in the system, then an optimal solution can be found through an integer linear programming problem.

In this exercise you will work with a simplified hypothetical region of 400 km² in area (Figure 21.1). The region has been divided into 100 square units of 2×2 km called Planning Units (PU).

Data are potentially available for 100 species from 4 taxonomic groups: 25 mammals, 25 birds, 25 amphibians and 25 fishes. You will design a set of reserves that represent all or a large subset of the species while minimizing the cost of implementing the network. We will assume that an area of 4 km² (2×2 km) can sustain viable populations of the species if no human-driven degradation of PU occurs. In other words, if the PU is selected and the reserve implemented, the populations inside the reserve are safe.

Note that all the species are not in all the PUs due to heterogeneity of the environment through the region. Thus, some sites contain few species and others contain many species. To do this exercise you first need to assemble all the data

Fig. 21.1 Study region divided into 100 planning units (PU).

and generate your first reserve network (Part 1). Then you have to choose the first PU to be included in the reserve, the one you will lose if you do not act quickly (Part 2). You may or may not have enough funding to collect information for all 100 species. Therefore, in Part 3 you will have to decide which taxonomic group alone provides a reserve network that is similar to the one using all the taxonomic groups together. In Part 4 you will explore two cost scenarios: In the first scenario every PU costs the same, and in the second the cost of each PU varies.

Objectives

The objective of this exercise is to introduce a basic technique to prioritize conservation at a regional scale (Margules & Pressey 2000). After this exercise you should be able to:

- Determine the data and structure required to set priorities (planning units × species matrix; cost of planning units; and vulnerability of planning units)
- Determine a network of reserves that minimizes the cost of implementation using linear programming while representing each species at least once
- Define priorities in terms of vulnerability
- Recognize the trade offs in using surrogates
- Assess the effect of cost of planning units
- Use readily available optimization software (Solver add-in in Excel or OpenOffice).

Procedures

Part 1 Setting the Problem

You are provided with the species that are occupying each PU (puspec.txt) at this book's website. You will import the data into MS Excel® to generate a matrix of PU and species (Figure 21.2). If you are new to MS Excel® do not worry; we provide instructions for using the Text Import Wizard and PivotTable Wizard at the same website.

The matrix represents a region divided in 5 planning units and with distributional data for 4 species. Notice that all planning units have 2 species, SP1 is only found in PU1, SP2 and SP3 are found in 2 planning units respectively and SP4 is in every planning unit. Remember that in this exercise you will be dealing with 100 PUs and 100 species and that real problems easily deal with more than 10 000 PUs and several hundred species.

Using the format in Figure 21.2 you will find the minimum combination of sites that represents all the species at least once. Therefore, you have a simple **conservation goal**: At least one population of each species must be in the reserve network. You may well be interested in conserving more than one population per species, or even all the known populations per species for some of them, but let's start with the simplest goal of protecting at least one population per species.

In this Part you will assume that each PU requires the same amount of resources to be managed for conservation; consequently minimizing the number of PUs involves minimizing the money spent to represent all the species.

Given that you want to minimize the overall cost of your reserve system, you have to create a variable representing that cost. In this case, a simple variable is the number of PUs. To state if each planning unit is considered or not in the reserve network you will create a binary variable. The variable will have a value of 1 if the PU is in the reserve network or 0 otherwise. Basically you have to add a new column after the last species column with the name Rx, or the name of your choice, in the heading. Please take a look at Figure 21.3. Remember, if $Rx = 1$ the PU is reserved; otherwise it is not, that is, $Rx = 0$. Since none of the PUs are in the reserve network yet, you can assign a value of 0 to all Rx. Keep in mind that the value you want to minimize in this part is the number of PUs being part of the reserve. In other words, you want to minimize the summation of the variable Rx. The summation of Rx constitutes the output of your Objective Function.

First, Figure 21.3A is the same matrix described in Figure 21.2 but now has a new variable (Rx) indicating the status of the planning unit: 1 = reserved, 0 = unreserved. In this example, 2 planning units, PU1 and PU4 are picked as

Planning Units	Species			
	SP1	SP2	SP3	SP4
PU1	1	0	0	1
PU2	0	1	0	1
PU3	0	0	1	1
PU4	0	1	0	1
PU5	0	0	1	1
Σ =	1	2	2	5

Fig. 21.2 Structure of the base data: A PU × Species matrix.

(a)

Planning Units	Species				Rx
	SP1	SP2	SP3	SP4	
PU1	1	0	0	1	1
PU2	0	1	0	1	0
PU3	0	0	1	1	0
PU4	0	1	0	1	1
PU5	0	0	1	1	0
Σ =	1	2	2	5	2

(b)

Planning Units	Species multiplied by corresponding Rx			
	SP1*Rx	SP2*Rx	SP3*Rx	SP4*Rx
PU1	1	0	0	1
PU2	0	0	0	0
PU3	0	0	0	0
PU4	0	1	0	1
PU5	0	0	0	0
Σ =	1	1	0	2

Fig. 21.3

part of a reserve network. Second, the multiplication of each value of 21.3a by the respective value of Rx is presented in 21.3b. Notice that the only PUs maintaining the original values are those considered in the reserve network, all other PUs have values of 0 for all species given that Rx is 0. Additionally, 21.3b presents the summation of the populations of each species represented in the reserve network. Also, notice that SP3 is not represented in the reserve network. If you want to represent SP3 you will have to reserve PU3 or PU5.

Now you have to use a second set of variables indicating the number of occurrences of each species in the reserve system you are proposing. To achieve this, you have to use the variables SP1 to SP100 and Rx to generate a new matrix (see Figure 21.3b to get an idea of the structure of the matrix). This matrix has the same structure as the PU × Species matrix in Figure 21.2, but the values are 0 if the value of variable Rx is 0 and are equal to the original value if the $Rx = 1$. In other words, you do not count the species in the unreserved PUs. You can create this new matrix by multiplying the value of the original matrix by the respective value of Rx (Figure 21.3).

Before continuing with the next step, check if your spreadsheet is working as you expect. You can start adding some PUs to your reserve network by modifying Rx from 0 to 1 and see how those changes are reflected in the species represented by the network. Try to change all Rx to 1 and explore what happens with the species represented by the reserve network; all of them must be represented. Then, you can start changing some Rx to 0 while keeping all species represented. What is the smallest number of PUs you can manually assign while keeping all species represented in the network at least once?

Now you are ready to solve the problem using MS Excel's Solver "add in." (Solver should be available under Tools; if not, activate by going to Tools/Add-Ins and select the option Solver Add-in.) It is highly recommended to explore the Help information under Solver Parameters (Figure 21.4). Set Target Cell is the Objective function (summation of Rx), allocated in the cell CY103 in this case. Equals To should be set to Min. since you are looking for the fewest PUs. Changing Cells should be set to the cell addresses of the Rx array: these are the cells the add in will change until some minimum number of PUs is found. In this case these are

Fig. 21.4 Example of basic parameters for Solver.

the cells above CY103, from CY2 to CY102. Subject to the Constraints sets the restrictions of the model. This problem requires two restrictions: (i) Rx is a binary variable (0/1) and (ii) the summation of those species present must be ≥ 1. The binary variables are the "changing cells" too; therefore in this example they range from CY2 to CY102. The variables indicating the presence of each species in the reserve network are the summation of the "second matrix" explained in Figure 21.3b. In this particular case the cells range from C106 to CX106. Be aware that you may have different cell addresses when you set your particular problem.

After entering the appropriate parameters you should be able to solve the problem. Note that your computer may require considerable time to solve the problem for 100 species and 100 PUs. However, you may want to check that the additional parameters of your model are correct before running to avoid potential problems. Go to Options; the options might look like those in Figure 21.5.

The result of this optimization process is that Rx values will change until an optimal solution is reached (or some failure occurs in the process that prevents it

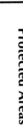

Fig. 21.5 Additional options for the model.

returning any result). *Rx* values similar to 1, like 0.999, indicate those PUs that are considered part of the an optimal solution for the reserve system.

At the end of this process, you should report the total number and ID of the PU's that minimize the cost of **representing all the species** (that is, the minimum combination of PUs that represents at least one PU for every species). Fill the map with the proposed set of PUs in Figure 21.6. Also consider repeating the experiment starting with a different set of values for *Rx*. If you started with all *Rx* being 0, you may try to replace all of them by 1 and then run Solver again. Additionally you may try other combinations, like adding *Rx* values of 0 and 1 at random. Would you expect the same result? Which PUs are the same, if any? In some optimization problems and techniques, the "seed" or first configuration of the system may influence the configuration of the optimal solution. Moreover, you may find several solutions achieving what you expect, a minimum number of PUs representing all species at least once. You may be interested to know that some commonly used "reserve algorithms" obtain several "optimal" solutions; those PUs that are always in the optimal solution are called "irreplaceable" with the remainder considered more "flexible" (see Margules & Pressey 2000 for more details).

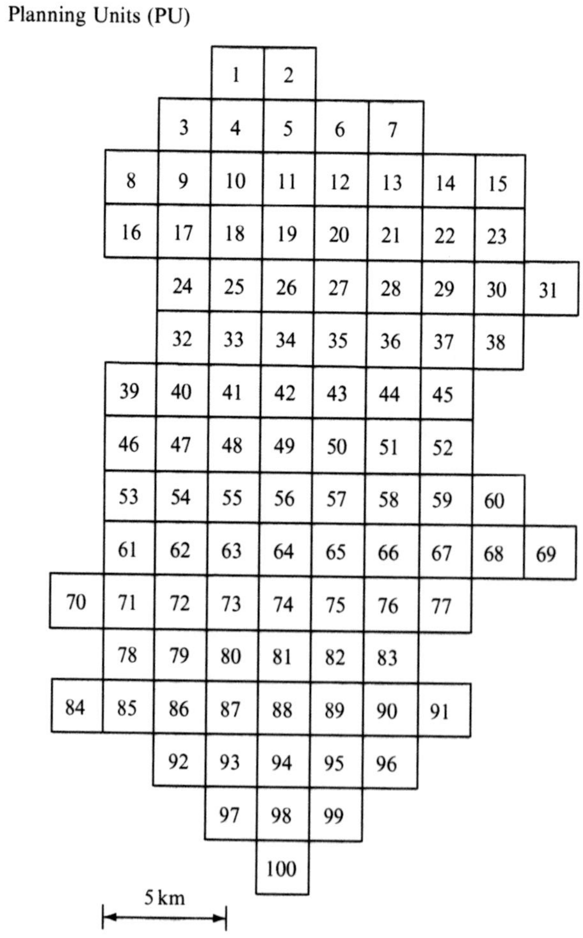

Planning Units (PU)

Fig. 21.6 Your proposed reserve network.

Part 2 Prioritizing the Reserve Schedule: Where to Go First?

Next you will decide which PUs are the highest priority for developing a conservation action plan by ranking the selected PUs using **vulnerability**. Vulnerability is the risk of an area being transformed in a way that no longer can support populations of native species (Margules & Pressey 2000).

In this exercise vulnerability is associated with a planned road that will cross the whole region (Figure 21.7), opening it up for oil and gas exploration. The road will degrade the forest up to 1 km away; the extent of ecosystem degradation (ha per PU) is quantified in the file "puinfo.txt" at this book's website.

Using the vulnerability values, state where you will start working on your conservation plan. Mark the selected PU in Figure 21.6, with a different pattern or color.

Part 3 Further Exploration: Using Partial Information About Species

In reality one is always dealing with incomplete knowledge about biodiversity, and thus surrogates such as subsets of species, species assemblages, and ecosystem types are often used as measures of biodiversity (Margules & Pressey 2000). In this part of the exercise you will determine the taxonomic group that best represents the other taxonomic groups. You will do this by comparing solutions using one taxonomic group against the solution using all the taxonomic groups. You will notice that this comparison is not required for this scenario, given that we already

Fig. 21.7 Study region divided in PUs along with the projected road.

have the complete set of species information. However, this comparison could allow you to be much more efficient in other, similar areas by focusing your survey efforts on one taxa. In those areas you might not have complete information on multiple taxonomic groups and it would be useful to have identified an indicator group first.

For this part of the exercise you will determine optimal reserves for each of the groups separately and then will compare the outcome with your optimal reserve (the one that used all taxonomic groups in Part 1). To obtain a solution for some specific taxonomic group you have to change the original configuration of the solver problem. The only change is in the constraint (see "Subject to the Constraints" in Figure 21.4). You have to click over the constraint you want to change, so it is selected, and then click on Change. You will be directed to a small popup window where you can edit the constraint the same way you entered it the first time in Part 1. There, instead of picking the entire range of species represented in the reserve network, you will select the taxonomic group of your choice (see Figure 21.8). Then you accept everything and solve again. The new result should represent every species of your taxonomic group but is not necessarily taking care of the rest of the groups. Copy the *Rx* column to some other place since you will use those values to compute Jaccard's coefficient and then repeat the process with another taxonomic group.

Use Jaccard's coefficient ($S_{i,j}$) to compare those solutions and report those values for every taxonomic group. Here is how you calculate Jaccard's coefficient:

$$S_{i,j} = a / (a + b + c)$$

where a is the number of cases where $i = 1$ and $j = 1$; b is the number of cases where $i=1$ and $j=0$; and c is the number of cases where $i = 0$ and $j = 1$.

Next you will find the PUs selected by the algorithm for all species, just mammals, just birds, just amphibians, or just fish (Table 21.1).

If you want to use Jaccard's coefficient to compare the solution representing all species and the one representing just mammals you have to determine a, b, and c for that specific comparison. The number of cases where both the solution representing all species and the one representing just mammals is 1 represents "a", which in this case is 1 (see PU 1). The number of cases where the solution representing all species is 1 and the one representing just mammals is 0 represents "b", in this case equals 4. Finally, adding the occasions when the solution representing all species is 0 and the one representing just mammals is 1 is called "c", and in this particular example is 1 (see PU 6).

Then Jaccard is $a / (a + b + c)$ or $1/(1 + 4 + 1)$ or 0.167.

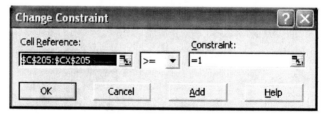

Fig. 21.8 Window used to change a single constraint.

Table 21.1

209

Protected Areas

Ch 21
Protected Areas

PU	Rx All	Rx Mammals	Rx Birds	Rx Amphibians	Rx Fish
1	1	1	1	1	1
2	0	0	0	0	0
3	0	0	0	0	0
4	1	0	1	0	0
5	1	0	0	1	0
6	0	1	0	1	0
7	0	0	0	0	0
8	1	0	0	0	1
9	0	0	0	0	1
10	1	0	0	1	0

You can repeat the procedure for birds, amphibians and fish. Jaccard's coefficient should be 0.4, 0.5, and 0.333, respectively.

Note that in the complete data set mammals are SP1 to SP25, birds are from SP26 to SP50, amphibians are from SP51 to SP75 and from SP76 to SP100 are fish species. Fill in the appropriate values in Table 21.2.

Part 4 Further Exploration: Using More Realistic Costs of Planning Units

The assumption that each planning unit costs the same makes the analysis simple but is not realistic. We may make this assumption in practice because sometimes we don't have any better information. In this part of the exercise you have some information that can provide a better estimation of the real costs involved. At this book's website, we provide a file "puinfo.txt" for each PU in which you have information about the length of the proposed road inside the PU and the presence or absence of a human settlement in the PU. Each PU costs US$1 m (million) to secure. If the PU contains a small settlement you have to add US$100 m. Also, the road building project has purchased a small strip of land to construct the road, so if you want to purchase the PU you will have to pay an additional US$ 10 m per km of road inside that PU. What is the cost of the entire area? As you might surmise it is the summation of the value of all the 100 planning units.

You have assigned a total cost to each PU using the information above and checked the value of the entire area. Now, the value of the PUs must be considered if they are part of the reserve network. To achieve this, you should add two additional columns to your problem setting as shown in Figure 21.9. The first

Table 21.2

Group	Jaccard coefficient of this taxon relative to all taxa
Mammals (SP1–SP25)	
Birds (SP26–SP50)	
Amphibians (SP51–SP75)	
Fish (SP76–SP100)	

(a)

Planning Units	Species				Rx	PUCost	Vx
	SP1	SP2	SP3	SP4			
PU1	1	0	0	1	*1*	*100.5*	*100.5*
PU2	0	1	0	1	*0*	*580.0*	*0*
PU3	0	0	1	1	*0*	*1304.1*	*0*
PU4	0	1	0	1	*1*	*10.0*	*10*
PU5	0	0	1	1	*0*	*55.3*	*0*
Σ =	1	2	2	5	**2**	**2049.9**	**110.5**

(b)

Planning Units	Species multiplied by corresponding Rx			
	SP1*Rx	SP2*Rx	SP3*Rx	SP4*Rx
PU1	1	0	0	1
PU2	0	0	0	0
PU3	0	0	0	0
PU4	0	1	0	1
PU5	0	0	0	0
Σ =	*1*	*1*	*0*	*2*

Fig. 21.9 Adding more variables to use actual cost of PU instead of same cost for all PUs.

column, adjacent to Rx is the corresponding value of each PU (PUCost), the second new column is the multiplication of Rx and the value of the PU (Vx). Therefore, if Rx is 1, Vx equals PUCost, otherwise, if Rx is 0, Vx is 0. Finally the summation of Vx should constitute the new objective function. As you are anticipating, the only change you have to make in solver settings is to replace the old objective function by the new one.

Two additional variables have been added to the original figure (Figure 21.3): PUCost or the total cost of the PU; and Vx or the multiplication of Rx and PUCost. The original objective function was the summation of Rx. The new objective function is the summation of Vx. All other parameters in solver should be the same.

You should report the total number of PUs as well as identify the PUs that minimize the cost of representing all the species. Highlight the proposed PUs in Figure 21.10. Are they different or the same as those in Part 1/Figure 21.6?

Expected Products

The following tasks will be completed:

- Identify the minimum set of PUs that represents all the species and map them in Figure 21.6.
- Identify the planning unit with the highest priority for conservation. Mark the PU in Figure 21.6 using a different pattern or color.
- Report the Jaccard's coefficients to compare the solutions based on all taxonomic targets versus those based on each single taxonomic group.
- Report the new reserve network that minimizes cost while representing all species and map your results on Figure 21.10
- Respond in a form indicated by your instructor to the Discussion questions below.

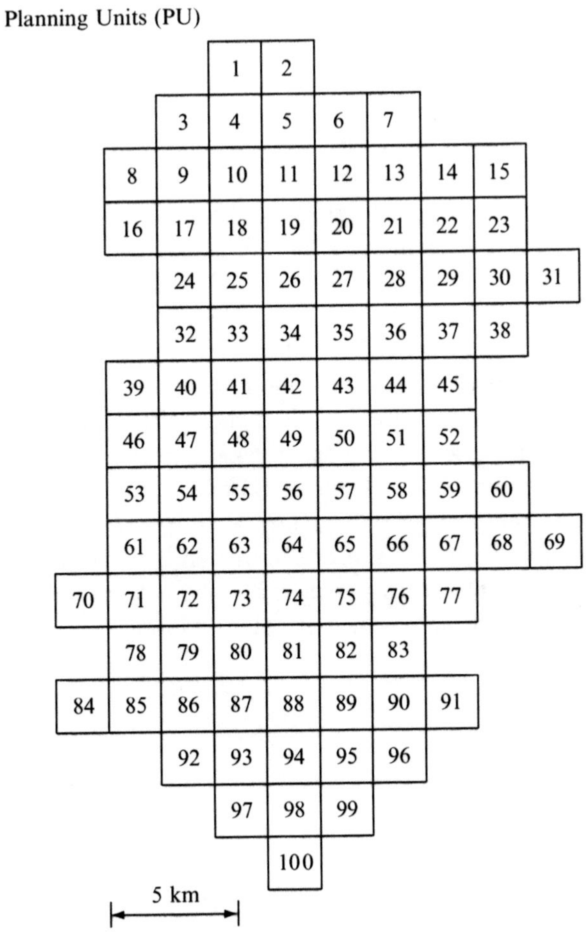

Fig. 21.10 The proposed reserve network.

Discussion

1 The reserve network in this exercise was selected with overall biodiversity "representativeness" in mind. What are some other components that should be taken into account when designing reserve networks?

2 Can you envision some surrogates different from the taxonomic groups? What about including coarse filter targets like land cover or environmental classifications?

3 Did you find differences in the solution using different objective functions? (e.g. minimizing number of planning units instead of total cost)? If so, do you think the cost outlined in Part 4 is enough? What about including potential management cost of the reserves? What else will you include in the objective function to make the optimization a more realistic exercise?

4 Think about all taxonomic groups, not just the four vertebrate classes used here; what features are likely to make a taxon a good surrogate for this kind of exercise?

5 How do you expect the cost of your solution will change if you use smaller planning units?

Making It Happen

Prioritization is required because conservation is inevitably conducted under constraints of time, land, or money. The process sketched in this exercise is being applied at several scales, from the entire globe to small watersheds. Regardless of the scale you are using, the key element to any prioritization exercise is defining explicit conservation goals. These conservation goals are related to targets (e.g. species, plant associations, geological types, etc.) whose spatial distribution is known or has been reliably estimated. To extend these ideas to where you live: What are potential targets in your region of interest? Are you able to define the spatial distribution of these targets? Are you and the stakeholders of your region able to agree about explicit conservation goals for each target? These are some of the important questions you will be confronted when you start with any prioritization process. Besides the technical aspects of setting priorities with algorithms, the importance of these methods resides in their flexibility and, above all, in their ability to coordinate people with contrasting perspectives about what must be done to represent a persistent sample of the biodiversity of a region.

Further Resources

Here is a list of software useful for conservation planning:

- Cluz web site (http://www.mosaic-conservation.org/cluz/index.html)
- Marxan web site: (http://www.ecology.uq.edu.au/index.html?page=27710)
- Sites web site: (http://www.biogeog.ucsb.edu/projects/tnc/toolbox.html)
- Resnet web site: (http://uts.cc.utexas.edu/~consbio/Cons/Labframeset.html).

CHAPTER 22

Island Biogeography: How Park Size and Condition Affect the Number of Species Protected

James P. Gibbs

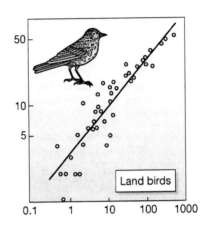

Islands are powerful, but ambiguous, metaphors. They convey images of safe havens, refuges that are buffered from the tumult of the world by a span of water. On the other hand, they are often seen as tiny, remote fragments, buffeted by disturbances. Conservation biologists, sensitive to the innumerable island species that have gone extinct during the last few hundred years, are more likely to see islands as fragile ecosystems, vulnerable to invasions by exotic species and a host of other problems. In a more general sense, many protected areas on land have essentially become islands of natural ecosystems in a sea of dominated by humans. Isolation is important but smallness, in particular, limits the number of species that can survive in an isolated protected area. The reasons are that, all other things being equal, a small area will have a narrower range of environments than a large area; uncommon species, especially those with large home ranges, are less likely to occur in a small area; and small areas have small populations that are inevitably more vulnerable to extinction.

This then raises the question: are larger areas indeed better at protecting more species? If so, there should be a positive statistical relationship between the size of a protected area and the number of species it hosts, with area alone explaining much of the variation in species richness. Conversely, should small protected areas be given a lower priority than larger ones? A related question is: just how effective are protected areas set aside expressly for conserving species versus other types of protected areas? In other words, do such areas protect more species per unit area and do they accumulate species more quickly with each increment of size than do protected areas that were not expressly created to protect biodiversity (e.g. national forests or recreational areas)?

You can examine all these questions if you have a list of protected areas, their status, and the number of species that they host by implementing the equation for the species-area relationship, which is:

$$R = cA^z$$

where R is the number of species, A is the size of the protected area, and c and z are empirical constants. If we first do a $\log_{10}$ transform of data for both sides of the equation we can get a straight line and use this to determine the constants. Here is the log-transform:

$$\log_{10} R = \log_{10} c + z \log_{10} A$$

These constants now correspond to the slope of a straight line (z) and the intercept of a straight line (c). Note that the intercept is not c, but the $\log_{10}$ of c.

By estimating these relationships with field data you can answer the questions posed above. More specifically, if larger protected areas are better at protecting more species than smaller reserves we should expect the relationship between size and species richness to be strongly correlated with the size of a protected area. We would also expect that protected areas set aside specifically for biodiversity protection should accumulate species per unit area faster than those set aside for other reasons, that is, the slope (z) should be greater. Similarly, one would expect protected areas of the same size that are set aside for biodiversity protection to host a larger number of species than those with non-biodiversity priorities, in other words, the intercepts (c) should be different.

In this exercise we address these questions by analyzing species lists from a large number of protected areas in the United States of varying size and protection status. We transform data on park size and species richness to test the strength of the relationship between protected area size and species richness and then estimate values of z and c to contrast protected areas with different biodiversity priorities. This amounts to an application of biogeography theory to an important question in conservation biology – the efficacy of protected areas for protecting wild species.

Objective

- To apply island biogeography theory to test basic assumptions about the role of nature reserves in protecting biodiversity.

Procedures

The Dataset

For this exercise we have compiled a dataset of plant and animal species richness from 160 protected areas in the United States, where inventory and monitoring of the biota has been ongoing for many decades and hence the data are reasonably reliable. The data were obtained from a variety of sources but mainly from Information Center for the Environment (ICE) at the University of California at Davis. The Center is a cooperative venture with the United States Man and the Biosphere program (US MAB), the National Biological Information Infrastructure, the US National Park Service, and the Biological Resources Discipline of the US Geological Survey. The database contains documented, taxonomically standardized species inventories of plants and animals reported from the world's protected areas. We have extracted for you information from federal protected areas on terrestrial portions of the contiguous states of the United States. To control for significant sources of heterogeneity we have excluded marine areas, islands, protected areas

Table 22.1 Example data used in this exercise.

Name	Priority	km²	Bird species	Name	Priority	km²	Bird species
Acadia National Park, ME	Biodiversity	190.19	346	Agate Fossil Beds National Monument, NE	Other	12.36	85
Arches National Park, UT	Biodiversity	296.96	153	Amistad National Recreation Area, TX	Other	236.75	206
Badlands National Park, SD	Biodiversity	982.42	219	Antietam National Battlefield, MD	Other	13.18	146
Beaver Creek Biosphere Reserve, AZ	Biodiversity	1113	111	Arkansas Post National Memorial, AR	Other	1.58	236
Big Bend Biosphere Reserve, TX	Biodiversity	2832.47	357	Aztec Ruins National Monument, NM	Other	1.29	37
…	…	…	…	…	…	…	…

in Alaska and Hawaii, and terrestrial areas protected by states, municipalities, and others. We have classified for you protected areas that have biodiversity protection as a primary objective (mainly national parks or biosphere reserves) versus all other categories of protected area status where biodiversity protection is often secondary to other mandates (primarily national monuments, national historical sites, national memorial sites, national recreation areas, national wildlife refuges, national battlefields, national military parks). We focus on birds and plants simply because these taxa are the best inventoried among all the groups for which data are available. A sampling of the data is provided in Table 22.1. The full set is available at: www.esf.edu/efb/gibbs/solving/

Transforming the Data

First you need to do a $\log_{10}(x)$ transform of data for both sides of the equation in order to get a straight line and use this to determine the constants. We recommend using the spreadsheet data you downloaded from the book's website to do so. For example, in the spreadsheet provided, go to cell F1 and enter Log10(area), and then go to cell F2 and enter: =LOG10(C2). This will provide a log transformation of the protected area size estimate (km²) currently residing in cell C2. Then copy the contents of F2 and paste them down the entire length of the column to transform the size estimates for all the remaining protected areas. Repeat this same process for the bird and plant species richness estimates. You will now have parallel columns of data ready for analysis that look much like Figure 22.1.

log_{10}

	A	B	C	D	E	F	G	H
1	Protected Area Name	Priority (Biodiversity versus Other)	Size (km2)	Bird species	Plant species	Log10(area)	log10(birds)	log10(plants)
2	Acadia National Park	Biodiversity	190.19	346	1119	2.28	2.54	3.05
3	Arches National Park	Biodiversity	296.96	153	392	2.47	2.18	2.59
4	Badlands National	Biodiversity	982.42	219	445	2.99	2.34	2.65
5	Beaver Creek Biosphere Reserve, AZ	Biodiversity	1113.00	111	78	3.05	2.05	1.89
6	Big Bend Biosphere	Biodiversity	2832.47	357	1049	3.45	2.55	3.02
7	Big Thicket Biosphere Reserve, TX	Biodiversity	342.17	174	1186	2.53	2.24	3.07
8	Biscayne National	Biodiversity	699.81	223	36	2.84	2.35	1.56
9	Bryce Canyon National	Biodiversity	145.02	189	316	2.16	2.28	2.50

Fig. 22.1

Plotting the Data

First, we recommend making plots of species richness (log10(x) transformed) by protected area size (also log10(x) transformed) and fitting a straight line to the data. Do this for

1 Birds on priority = "biodiversity" areas
2 Birds on priority = "other" areas
3 Plants on priority = "biodiversity" areas
4 Plants on priority = "other" areas.

Plotting these relationships on the same graph will give you a good sense of general patterns and facilitate comparisons. Plotting data in this manner is always a good idea prior to beginning statistical analyses. We have provided the plot below (Figure 22.2) but with its legend dropped for you to interpret based on your own regression analysis (or those presented below in Table 22.2).

Performing the Regression Analysis

Next, estimate the "fit" and parameters of the linear model. To do this you need to regress log(R) versus log(A). The regression coefficients will give you the parameters you need: z is the slope, and c is 10 raised to the power of the intercept.

You will need access to some statistical package to calculate these estimates for such a large dataset. There are many options; we suggest using the free PopTools add-in to MS Excel (www.cse.csiro.au/poptools/) (but you or your instructor can likely identify many other options for performing least-squares linear regression as either a stand-alone piece of software or as an internet-based application). If you do use PopTools you can find the easy-to-use regression function under PopTools/Extra Stats/Regression. Remember that log_{10}(area) is always the independent variable (X) and log_{10} plant or log_{10} bird species richness is always the dependent variable (Y). Perform these regressions – they should match the following summary from PopTools regression procedure in Table 22.2.

Interpretation

There are quite a few parameters to sort through here, so let's work through them.

Estimate

This is the estimate of c (log_{10} transformed) or z. The standard deviation indicates the variability about the estimate. The UCL (upper confidence limit) and LCL

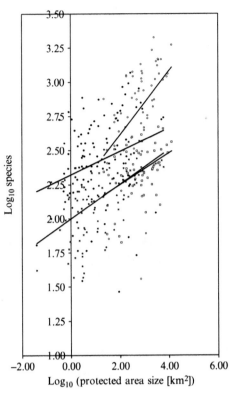

Fig. 22.2 General relationships between extent of protected area and species richness.

Table 22.2 Regression analysis of plant and bird species richness in relation to protected area size and condition.

	Parameter	Plants in biodiversity priority areas	Plants in "other" areas	Birds in biodiversity priority areas	Birds in "other" areas
	Sample size	57	102	57	102
$\log_{10}(c)$ →	Estimate	2.145	2.321	2.019	1.998
	Standard deviation	0.175	0.048	0.096	0.030
	UCL (95%)	2.191	2.330	2.044	2.004
	LCL (95%)	2.100	2.312	1.994	1.992
	Student t	12.269	48.596	20.991	66.384
	Prob(>|t|)	0.000	0.000	0.000	0.000
z →	Estimate	0.235	0.086	0.116	0.127
	Standard deviation	0.061	0.030	0.033	0.019
	UCL (95%)	0.250	0.092	0.124	0.131
	LCL (95%)	0.219	0.080	0.107	0.123
	Student t	3.876	2.826	3.476	6.635
	Prob(>|t|)	0.000	0.006	0.001	0.000
	R^2(adj)	0.200	0.064	0.165	0.297
	F	15.027	7.987	12.081	44.029
	Prob(>F)	0.000	0.006	0.001	0.000

is the relationship statistically significant?

(lower confidence limit) estimate the bounds within which we are relatively certain (at a confidence level of 95%) that the true values occur. Note that the UCL and LCL provide a convenient way to contrast estimates between taxa and among categories of protected areas. If the UCL of one group is less than the LCL of another then the parameter estimates differ.

Student t and Prob($>|t|$)

Together these refer to the probability that the estimate equals zero. As you can see, there are very small probabilities that these estimates equal zero; in other words, they likely do not.

R^2(adj)

This provides some measure of fit to the model. High values ($\cong 1$) indicate that much of the variation in species richness is explained by area alone. Low values ($\cong 0$) suggest that none is. The actual values suggest that size of a protected areas provides low to moderate levels of "fit" for variation in richness of birds and plants.

F and Prob($>$F):

These indicate how well the regression models perform overall. If these probabilities are less than 0.05, then we can assume that they perform "significantly" better than random in terms of explaining variation in species richness.

The last manipulation to make is to re-calculate the confidence intervals about the intercept. Recall that the intercept is not c, but the $\log_{10}$ of c. So we need to take 10 to the power of the estimate of c to estimate the actual intercept in terms of species. The estimates are provided in Table 22.3.

A Final Exploration

Inventory is expensive and we can only afford to do it for select taxa. Can we be conservative and assume that diversity covaries among groups? In other words, can we just inventory one group and assumes it reflects variation in other groups? You have the data to evaluate this important issue. We suggest regressing, for example, bird species richness on plant species richness *and* protected area size. Why? You must control for area alone before looking at covariation between birds and plants. Consider doing this analysis to explore whether inventory of one group can be a good surrogate for another.

Table 22.3 Confidence intervals about the estimates of the intercept for bird and plant diversity in protected areas of different status.

	Plants in biodiversity priority areas	Plants in "other" areas	Birds in biodiversity priority areas	Birds in "other" areas
UCL (95%)	155.1	213.9	110.6	100.9
LCL (95%)	125.8	205.0	98.6	98.2

- Log-log plots of: (i) birds on priority = biodiversity areas, (ii) birds on priority = other areas, (iii) plants on priority = biodiversity areas, and (iv) plants on priority = other areas.
- Your instructor may require you to perform the regression analysis and provide the details separately from what has been provided to you here.
- Responses based on the analysis to the discussion questions posed below.

Discussion

1. Is protected area size alone a good predictor of species richness? Is it better for birds or plants? What might explain the remaining variation?
2. Are there generally more plant or bird species per unit area? Explain this in terms of the estimated c-values.
3. Do protected areas with express mandates tend to support and accumulate more species per unit area than those without such mandates? Why do you think the rate of plant diversity suddenly increases at larger, better protected sites? Explain this in terms of c- and z-values for birds and plants separately.
4. Are protected areas with express mandates for conserving biodiversity larger or smaller than those without such mandates? Inspect the graph or do some quick calculations with the "=average()" function in the spreadsheet. This leads to the next question: Are these types of protected areas likely hosting the same suites of species? What suites of species are likely represented in the higher counts for the parks and protected areas? Why do you think the rate of plant diversity abruptly increases at larger, better protected sites?
5. Can we conclude that areas with express mandates for conserving biodiversity are better at protecting it than areas without such mandates? Explain your answer in terms of the regression outputs for both plants and animals.
6. If you pursued the last question of surrogacy of bird diversity for plant diversity, what did you conclude?

Making It Happen

Many of the relationships observed in this exercise will hold up on a smaller spatial scale in your local area. It could be very useful to construct similar curves for species of interest to you in patches of natural ecosystems in your area. It would require much effort, but it can be done across many, variably-sized patches in several ecosystem types that support distinctive communities, such as forests, wetlands, lakes, and grasslands. You can then use these numerical, species-area relationships to determine roughly how big a patch needs to be to support varying fractions of the total species in a community. Lastly, you can compare the areas of designated parklands in your region to see how adequate they might be for supporting various communities. These data potentially provide a powerful means for justifying expansion of protected areas in areas where parklands are small and isolated.

Further Resources

MacArthur and Wilson (1967) is the original source for these ideas and Quammen (1997) is a readable popular treatment of them. For an intriguing extension of these ideas to rates of species loss in US national parks, see Newmark (1987). You are encouraged to consult with the International Biogeography Society, which has a strong interest in conservation biogeography, for further information: http://www.biogeography.org/index.htm. On a related note, this group has compiled a large set of sample datasets from all manner of island-like situations that are amenable to the kinds of analyses described here: http://128.196.198.6/~michaelweiser/island/.

CHAPTER 23

GIS for Conservation: Mapping and Analyzing Distributions of Wild Potato Species for Reserve Design

Robert J. Hijmans

Comparing the desolate expanses of the Antarctic icecap to the vibrant, teeming forests of the equator is an extreme way of highlighting a fundamental biological fact: life is very unevenly spread over our planet. Understanding what maintains and threatens these patterns, in particular trying to discern places where assemblages of species are especially diverse and jeopardized by human activities, is a key activity for conservation biologists. That means analyzing masses of complex biological and geographic data. Fortunately, computers are providing us with some extraordinary tools for this, and are thus increasingly being used to help us set conservation priorities rationally.

In this exercise, you will investigate the geographic distribution of a group of species, and try to develop some strategies for its conservation. You will learn how to do this with a simple GIS program. GIS stands for "Geographic Information System," a fancy term for software for creating, displaying, managing and analyzing geographic data. You will use data for *Solanum* sect. *Petota*, over 100 wild species related to the cultivated potato, *Solanum tuberosum*, that occur in the Americas. Analyses similar to what you will do in the exercise are described in Hijmans & Spooner (2001), which can serve as a useful companion document as you work through the exercise and address the discussion questions.

Objectives

- To learn how to import taxon point locality data to a GIS program and map species distributions
- To develop skills in making maps and analyzing patterns of diversity
- To perform analyses and make recommendations about reserve selection.

Procedures

First of all, download and install on your computer a free GIS program called DIVA-GIS from www.diva-gis.org. Download the datasets for this exercise at this book's website. It would not hurt to read chapters 1 and 6 of the manual, and do the tutorial, which you will also find on the DIVA-GIS website. This software and others like it provides you with a generic tool to ask fundamental questions about the distribution of any taxon and to develop a rationale for basing conservation decisions.

Import Data

Create a folder called (e.g.) "divawp." Download the data for this exercise from the book's website, put them in the new folder and unzip the files. Open the file "wildpot.txt" with a spreadsheet program such as Microsoft Excel to convert the data in degrees and decimal minutes to decimal degrees (see the Figure 23.1 and the Geographic Coordinates box below as well as, if needed, Section 1.5 of the DIVA-GIS manual).

To do the conversion, insert new columns (labeled LAT and LON) in your spreadsheet just after LATH and after LONH. To calculate longitude in each cell, use a formula like this:

$$= -1 * (D2 + E2/60 + F2/3600)$$

As these plants are restricted to the Americas (Western hemisphere), all longitudes have a negative sign. This is not the case for latitude. To calculate latitude, we apply the negative sign only when the hemisphere equals S. Use a formula like this:

$$= IF\ (L2 = \text{``S''}, -1 * (I2 + J2/60 + K2/3600), I2 + J2/60 + K2/3600)$$

	D	E	F	G	H	I	J	K	L	M	N
1	LongD	LongM	LongS	LongH	LON	LatD	LatM	LatS	LatH	LAT	SPEC
2	65	45	0	W	-65.7500	22	8	0	S	-22.1333	S. ac
3	66	6	0	W	-66.1000	21	53	0	S	-21.8833	S. ac
4	65	5	0	W	-65.0833	22	16	0	S	-22.2667	S. ac
5	66	15	0	W	-66.2500	22	32	0	S	-22.5333	S. ac
6	66	12	0	W	= -1 * (D6+E6/60+F6/3600)	22	30	0	S	=IF(L6="S",-1* (I6+J6/60+K6/3600), I6+J6/60+K8/3600)	S. ac
7	66	12	0	W	-66.2000	22	28	0	S	-22.4667	S. ac
8	66	7	0	W	-66.1167	21	53	0	S	-21.8833	S. ac
9	66	6	0	W	-66.1000	21	53	0	S	-21.8833	S. ac
10	65	3	0	W	-65.0500	22	12	0	S	-22.2000	S. ac
11	65	0	0	W	-65.0000	22	9	0	S	-22.1500	S. ac
12	66	11	0	W	-66.1833	21	53	0	S	-21.8833	S. ac
13	66	11	0	W	-66.1833	21	53	0	S	-21.8833	S. ac
14	66	7	0	W	-66.1167	21	53	0	S	-21.8833	S. ac
15	65	52	0	W	-65.8667	25	8	0	S	-25.1333	S. ac
16	66	7	0	W	-66.1167	21	53	0	S	-21.8833	S. ac
17	66	3	0	W	-66.0500	17	14	0	S	-17.2333	S. ac

Fig. 23.1 Using Microsoft Excel to convert coordinate data in degrees, minutes, and seconds to decimal degrees.

Geographic Coordinates

Locations on Earth can be formally described using degrees of longitude and latitude. A location can be between 180 °W and 180 °E (longitude), and between 90 °N and 90 °S (latitude). Degrees can be subdivided using a sexagesimal system (a numerical system with 60 as the basic number) of minutes and seconds, just as we do with hours. For example, a latitude can be 12° 34′ 15″S (12 degrees, 34 minutes, 15 seconds, Southern hemisphere). In computers, latitude and longitude are more easily expressed as decimal numbers, with the sign indicating the hemisphere (+ = N or E, − = S or W) (e.g. −12.57083 for 12° 34′ 15″S). To convert longitude and latitude in degrees, minutes and seconds to decimal degrees, use the following formula:

$$DC = b \cdot \left(d + \frac{m}{60} + \frac{s}{3600}\right)$$

Where DC is the decimal coordinate; d is the degrees (°), m the minutes (′), and s the seconds (″) of the sexagesimal system, h = 1 for the Northern and Eastern hemispheres and −1 for the Southern and Western hemispheres. For example, 30° 30′ 0″S = −30.500 and 30° 15′ 55″N = 30.265.

Save the file first in Excel (or whatever spreadsheet you are using) format and then save it again as delimited text format (but with a new file name). Use tab-delimited; other formats, such as comma delimited, are OK but more prone to error. Import the new text file to DIVA-GIS (see below). If something has gone wrong (and if often does), you can return to the saved spreadsheet file, which will still have the formulas that are not saved in a text file, make the required changes, and then resave as delimited text format again.

Use the text file of coordinates to make a "shapefile" called "wild potatoes." Do this using Data >Import points to shapefile >From text file. Save the output file in the "divawp" folder and add it to the map. Make sure you choose the fields you just created for latitude and longitude (LAT and LON in the screenshots shown here).

A shapefile is a commonly used GIS file type; it is in fact a combination of at least three files that have filename extensions .shp, .shx and .dbf. See the DIVA-GIS manual for more details about this and other file formats.

Add the "pt_countries" shapefile to the map (using Layer >Add). Make sure that your map now looks like Figure 23.2.

Summarize by Country

Let's first summarize the data by country. Make the "wild potatoes" shapefile the active layer (click on it once in the Table of Contents so that it comes up) and click on Analysis >Point to Polygon. On the main tab, add the shapefile of countries in the Define Shape of Polygon box and select an output filename. On the Select Field tab select the Species field, return to the main (Options) tab and press Apply.

The result is a new countries layer. Make this layer visible and the other two layers invisible. Double click on the new layer and change its legend attributes.

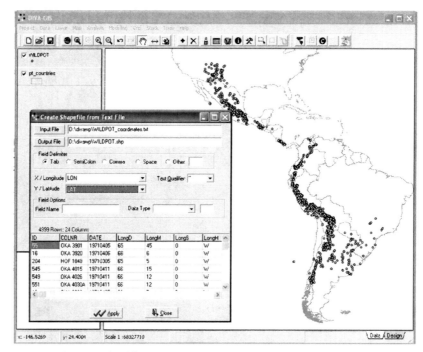

Fig. 23.2 Distribution of wild potatoes.

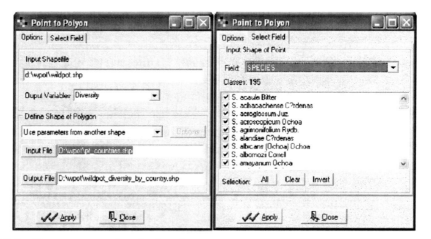

Fig. 23.3 How to summarize the data by country.

First, go to the Single tab, double click on the symbol and change its style to Solid fill. Then go to the Classes tab, select the SPP field, 6 classes, and reset the legend. SPP stands for the number of distinct species. The other fields are the number of observations and various diversity indices (see the DIVA-GIS manual). After that, change the values of the different classes to get something like Figure 23.4. This map shows that Peru is the country with the most potato species, followed by Bolivia and Mexico (which country has more?), and then by Argentina, Colombia and Ecuador.

If you open the table associated with the shapefile (Layer > Table), you will see that there are records for many countries that have no potatoes at all.

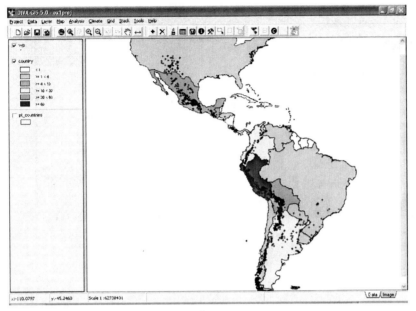

Fig. 23.4 Depicting the number of species by country.

To simplify our table, we first select the records with more than 0 observations (OBS >0), using the Layer >Select Records and the Select by Query tab. After entering the query, all countries with potatoes will be colored yellow on the map (they are "selected") (Figure 23.5). Now save the selection to a new shapefile (Data >Selection to new shapefile).

Now open the table showing the data for the selected countries. The table should look like Figure 23.6. It shows the number of observations (OBS), number of species (SPP) and four diversity indices for each country (see "Measuring diversity" box).

The data on numbers of species can be useful for some applications, but probably not very useful, because countries have such different sizes and shapes, and because most are very large. It is in most cases better to compare areas of equal area, such as square cells in grids, and that is what we will do next. But first use Project >Save to save the current project as, e.g., "exercise1A."

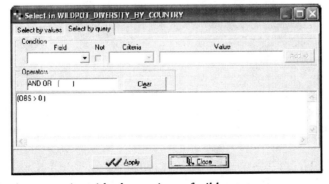

Fig. 23.5 Selecting countries with observations of wild potatoes.

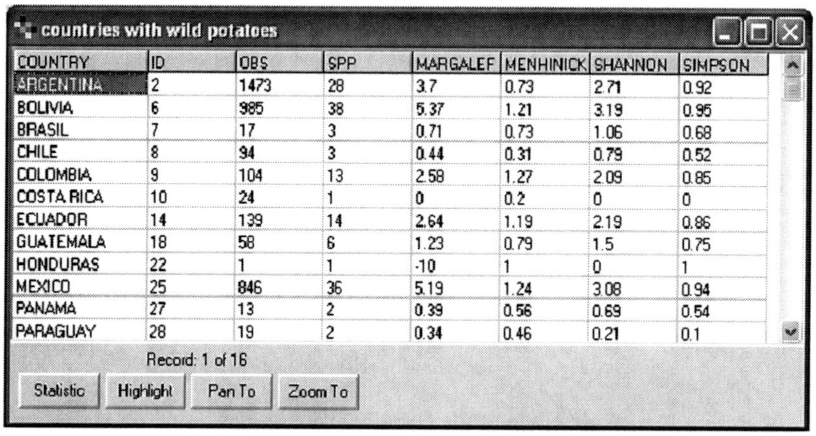

COUNTRY	ID	OBS	SPP	MARGALEF	MENHINICK	SHANNON	SIMPSON
ARGENTINA	2	1473	28	3.7	0.73	2.71	0.92
BOLIVIA	6	985	38	5.37	1.21	3.19	0.95
BRASIL	7	17	3	0.71	0.73	1.06	0.68
CHILE	8	94	3	0.44	0.31	0.79	0.52
COLOMBIA	9	104	13	2.58	1.27	2.09	0.85
COSTA RICA	10	24	1	0	0.2	0	0
ECUADOR	14	139	14	2.64	1.19	2.19	0.86
GUATEMALA	18	58	6	1.23	0.79	1.5	0.75
HONDURAS	22	1	1	-10	1	0	1
MEXICO	25	846	36	5.19	1.24	3.08	0.94
PANAMA	27	13	2	0.39	0.56	0.69	0.54
PARAGUAY	28	19	2	0.34	0.46	0.21	0.1

Fig. 23.6 The number of observations, number of species and four diversity indices for each country.

Project the Data

Projecting mapped data appropriately is an important issue to be familiar with (see Map Projections box). For projecting our wild potato data, remove all layers from

> **Map Projections**
>
> While spatial data with geographic coordinates (latitude, longitude) can be used to make maps, as we have seen, the maps produced are somewhat problematic because we are plotting coordinates on a sphere (the globe) to a flat surface. This causes distortions of the shapes and sizes of features on the map and of directions, among other things. To deal with these issues, a number of "map projections" have been developed. There is no "perfect projection" because you simply cannot perfectly represent a three dimensional sphere in two dimensions. But there typically is a good projection for your needs and study area.
>
> In ecology and conservation biology, study areas are often divided into grid cells. It can be important that all grid cells have equal area. This is because otherwise the comparison between grid cells would have to take into account the difference in grid cell size (e.g. larger cells will tend to have more species than smaller cells just because they are bigger). If latitude/longitude data were used, areas of say 1 square degree would get smaller moving away from the equator: think of the meridians (vertical lines) on the globe getting closer to each other as you go towards the poles.
>
> For relatively small areas, the UTM (Universal Transverse Mercator) map projection is frequently used to obtain equal area grids. In this exercise, we use a projection that can be used for a whole hemisphere: the Lambert Equal Area Azimuthal projection. Before you project your data, you must choose a map origin for your data. This should be somewhere in the center of your area of interest (distribution of points in this case), to minimize the distance (and hence distortion) from any point to the origin.

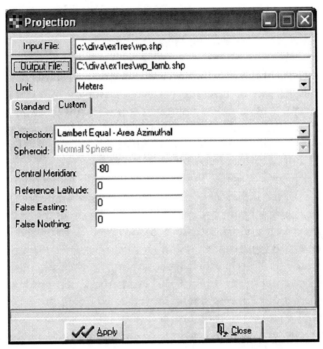

Fig. 23.7 Projecting the wild potato and the original country shapefiles using Tools/Projection.

the map, except the wild potato and the original country shapefile. To remove layers, use Layer >Remove layer or press the × on the shortcut menu bar.

Project the wild potato and the original country shapefiles using Tools >Projection. Choose the Lambert Equal Area Azimuthal (equatorial) projection (see textbox above and Figure 23.7). On the Custom tab, change the central meridian to −80 and keep the latitude at 0. Save the files with filenames such as "wild potatoes lambert" and "countries lambert." Press Apply. Be patient...

Make all layers visible. If you zoom to the maximum extent, you will see the projected data. Note that the shape of the countries is much more similar to their shape on a globe than before projecting. In the bottom left corner you can see that the coordinate system has changed. You will likely see very large numbers now: these express the distance from the origin (−80, 0 degrees) in meters. If you make the projected potato file invisible, you will see that the unprojected data is still present on the map, near the origin of the projected data. Zoom in to that area. Clearly, one cannot combine projected and unprojected data (or data in two different projections) on a single map.

Now go to Map >Properties and change the projection to "Other" and the units to meters. This will allow displaying a correct scale on the map.

Save the project to Exercise 1B.

Mapping Diversity Patterns on a Grid

Now we will determine the distribution of wild potato species richness using a grid (Figure 23.8). This can be done using the point-to-polygon option that we used before (by defining square or hexagonal polygons), but in most cases it is easier and more appropriate to use Analysis >Point to grid.

Measuring Diversity

A first step in the analysis of the distribution and conservation status of groups of species is often a description of diversity patterns. Diversity is not an entirely straightforward concept. The simplest approach to measuring it is determining "richness" by counting the number of distinct entities (typically species) in an area. There are also diversity indices that account for evenness, the relative frequency of species. For example, given a number of species present, such indices give a higher score to an area in which the species are equally abundant, compared to an area where most of the individuals encountered are of a single species, while the remaining species each account for a small proportion of all individuals.

A pervasive problem with most measures of diversity is their sensitivity to sample size: as sample size increases, the number of species goes up. Approaches to account for this include "rarefaction" (subsampling so that all sites have the same number of observations), and the use of "estimators" that predict the true number of species (at infinite sample size) (Colwell & Coddington, 1994).

Others have worked on incorporating taxonomic and phylogenetic information (Faith 2002). Their concern is that a group of species that are evolutionarily closely related should be regarded as less diverse than an assemblage of the same number of very distantly related species, particularly in the context of conservation planning. Magurran (2004) provides a detailed, but very accessible, description of approaches in quantifying biodiversity.

Fig. 23.8 Defining a grid for analysis.

Select "Richness" and "Number of different classes." Under Define Grid, create a new grid, press the Options button to set some parameters. In the Options window, set the X and Y resolution to 100 000 (as the projection is in meters this means that the cells will be 100 by 100 km). Use the default option (Simple) for the point-to-grid procedure.

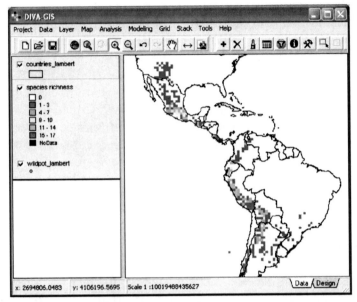

Fig. 23.9 Wild potato species richness on a grid.

Choose an output filename (e.g., "species richness"). On the Parameters tab, select the Species field. Press Apply (Figure 23.9).

Now make a grid of the number of observations, instead of species richness. To do this, do the same as for species richness, but change the second option under Output Variable on the Point to Grid window. Also, define the new grid with the option "Use parameters from another grid," and select the species richness grid you just made. In this way, you assure that you use exactly the same grid (in terms of cell size, number of rows and columns, and geographic origin) for the different analyses.

Compare the two grids with Analysis >Regression. Is the number of species in a cell a function of the number of observations? (See Figure 23.10.) As you can see, there is a relation between the number of observations and the number of species.

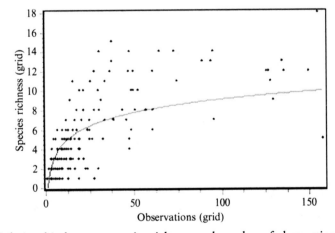

Fig. 23.10 Relationship between species richness and number of observations per grid cell.

Whether the number of species is, therefore, an artifact of collecting bias is another matter. The problem is that this association will always exist. When there are only a few species in an area, collectors will not continue to go there to increase the number of (redundant) observations.

Here we use a statistical measure — called R^2 — of the extent to which the values of one variable depend on those of another. Its value varies between 0 (the two variables are entirely independent) and 1 (entirely dependent). In this case, the coefficient of determination (R^2) is not very high (0.67) for this type of relationship. Also, there is a clear pattern in species richness maps. They are not characterized by sudden random changes in richness between adjacent areas: the data have **spatial autocorrelation**. Spatial autocorrelation means that cells near each other tend to have similar values. While this can be a problem in statistical analysis (if data points are assumed to be independent) in this case it is probably a good thing, because at least it is what you would expect to be the true situation. Diversity typically changes gradually from low to high, and not in a very abrupt and random pattern.

As stated in the measuring diversity textbox above, additional ways to look at this "collector-bias" problem include the use of **rarefaction** and **richness estimators**. Both methods are implemented in DIVA-GIS, under Analysis >Point to grid >Estimators of richness.

There are often gradients of species richness along latitude and altitude. An easy way to investigate a latitudinal gradient in species richness is by making a grid of a single column. Here we use the original, unprojected, shapefile. In the grid options window (in the Point to Grid dialog), use Adjust with Resolution and set the number of columns to 1.

After pressing OK, the program will give you a message about unequal resolutions, but you can ignore that as it is just what we want here. Create the species richness grid, which will look rather peculiar (a collection of horizontal bands), and then use Grid >Transect to create a figure like the one below. The distribution of species richness has two peaks. What might explain the low species richness between $-5°$ and $15°$? We think that this is because there are not many areas that are like the relatively dry and cold areas of the central Andes where wild potatoes are most common.

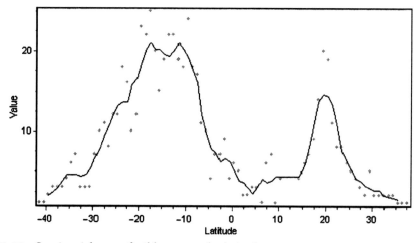

Fig. 23.11 Species richness of wild potatoes by latitude.

Now try an alternative way to make this graph. Export the data in the grid to a text file (Data >Export gridfile), import that file into MS Excel® or similar spreadsheet application, and make a similar graph using that software's graphics capabilities.

Reserve Selection

Let's now have a look at the implications of all these patterns in the context of designing a reserve system for these species. We would like to know how many protected areas we would need, and where these would have to be, to efficiently conserve these species in the wild.

Use the projected data again. Go to the Analysis >Point to grid >Reserve selection tool, and use the same 100-km resolution grids as before. On the parameters tab, select the species field and otherwise use the default values, and uncheck the "Maximum number of iterations," so that the algorithm won't stop until all species have been included in a reserve. The algorithm tries to find the minimum number of cells that are needed to include each species at least once. Although you supply only one output file name, three files will be produced. The file sequence will show the order in which different cells were selected (1..n, with the cell with the largest number of species first), another file shows the additional number of species that were added by that cell to the total in all selected cells (starting with a high number, and decreasing to at least 1), and the last file shows the number of species in each of the selected cells. How many cells are necessary to collect all species (at this resolution)? What is the median number of "additional species" that a selected cell contributes?

The "Rebelo" reserve selection algorithm used here is a very basic, but illustrative, reserve selection algorithm. There are a number of more sophisticated algorithms available elsewhere, such as C-Plan and Marxan (see Further Resources), but this one works fine for illustrative purposes.

Expected Products

- Map of distribution of wild potatoes
- Map of number of species per country
- Map of species richness by 100-km square grid cell
- Graph of latitudinal gradient in species richness
- Map of location of reserves and the estimate of their contribution to conservation
- Responses in a form indicated by your instructor to the Discussion questions below.

Discussion

1 What are the relative advantages of mapping species richness by country and by square grids?
2 Is 100 km square a sensible size for grid cells?
3 How else could these species be conserved apart from in reserves? And what kinds of analyses could be done in support of those activities?
4 What might be another taxon and conservation scenario that would benefit from this type of analysis? Would sufficient distributional data be available?

Making It Happen

You can refine this exercise by removing the apparently common species (e.g. those with many observations and/or a large range size). You could also first overlay the distribution of wild potatoes with a polygon layer of nature reserves (such a layer can be downloaded from the internet: can you find it?) and again remove the species that already appear to be protected, or at least present in a protected area. These two steps would allow you to focus on the species that are most in need of protection. You could also look at land-cover maps to identify areas that appear to be most prone to land-use conversion to agriculture or urbanization. Instead of point distribution data, you could use polygon (areas) range maps or ranges derived from species distribution models.

Further Resources

For those interested in exploring these issues further we recommend the following resources:

- Primary species occurrence data: www.gbif.org
- Georeferencing of localities: www.biogeomancer.org and Wieczorek et al. (2004).
- Diversity analyses: Colwell and Coddington (1994), Faith (2002), Gaston (2000), Orme et al., 2005, Graham and Hijmans (2006).
- Range maps: www.natureserve.org/getData/animalData.jsp
- Reserve selection: www.ecology.uq.edu.au/marxan.htm
- Geographic data: www.diva-gis.org, https://zulu.ssc.nasa.gov/mrsid/

CHAPTER 24

Global Change: Will a Cold-Adapted Frog Survive in a Warmer World?

Viorel Popescu and James P. Gibbs

Polar bears clinging to a shrinking ice-flow and tropical islands inundated by storm waves are just two of the many images that have become the face of global climate change. A diverse array of people is scrambling to cope with this issue: biogeochemists study the global distribution of carbon, engineers develop new "carbon-light" technologies, and policy specialists seek solutions that are economically and socially acceptable. Conservation biologists are particularly focused upon understanding what climate change will mean for the distribution of species and ecosystems. Amphibians are of particular concern because their physiological processes are closely associated with ambient temperature and moisture levels. Will climate change radically alter distributional patterns for amphibians? Or will the predicted changes in climate occur within the range of environmental conditions these species already experience across their current range? These are important but rarely asked questions.

How are these types of problem approached? First, we must introduce a few concepts. Scientists looked at how specific climate parameters (or derivatives of them) influence the distribution of animal and plant species by constructing "climate spaces" or "bioclimatic envelopes." Both these terms refer to the full range of climatic tolerances of a given species. Estimating currently occupied climate "spaces" or "envelopes" is useful for predicting both the current areas of suitable climate where a species has not been reported but it might occur (see Exercise 25 on wild peanuts) or future distributions based on predictions from climate change models. Here we examine the likely future range of a cold-adapted frog in relation to predicted warming of its range. Does it already span a sufficient amount of climate space to accommodate the changes predicted to occur? If not, will the forecasted change overwhelm it?

Objective

- To develop technical expertise in as well as a conceptual understanding of modeling species distributions in relation to changing climatic conditions.

Procedures

Study organism

For this exercise we focus on the mink frog (*Rana septentrionalis*), a pond frog of northern latitudes in North America. Mink frogs occur in cold northern lakes, ponds, and streams with open, shallow waters dotted with lily pads and fringed with emergent vegetation. They breed, feed, and hibernate exclusively in water. Eggs are laid in globular masses of up to 500 eggs that swell in aggregate to 7.5 to 15 cm in diameter. Egg masses are initially attached to submerged vegetation but often drop to the bottom. Quite common in its restricted habitats, the mink frog is known as "the frog of the north" because it has the most northern southerly limit of any frog in North America. It does not reach south of latitude 43 °N anywhere throughout its range.

Why is this species a good candidate for this exercise? Notably, the mink frog's globular egg mass and preference for permanent bodies of water are thought to establish the limits of its range. Whereas most pond-breeding frogs lay their eggs singly, in strings, or in spreading mats, such that oxygen in the water can easily diffuse into each egg, the globular nature of the mink frog egg mass impedes the diffusion process of oxygen to the eggs in the interior of the mass, which often die in the more poorly oxygenated and warmer waters found farther south. Colder, more northerly waters are usually sufficiently oxygenated to diffuse throughout the egg mass and sustain all the embryos. It is not clear why the globular egg mass of the mink frog evolved in the first place, but it may now constrain the species' extension into more southern regions of our state. Mink frogs are vulnerable to desiccation and rarely occur on land except during heavy rains when dispersal movements can be extensive. The species' biology therefore has a direct association with temperature and rainfall, making it an excellent candidate for an assessment of this type.

Information Base

Here we focus on a part of the species range, in New York State. We have chosen this region because it encompasses the southern boundary of the species range and because, thanks to the Amphibian and Reptile Atlas of New York State, we have extensive information on the species' current distribution (Figure 24.1).

You are provided three datasets in map form to work with:

1 Current known occurrence of the mink frog in New York (Figure 24.1)
2 Mean temperatures in New York State during July, an index of breeding season temperatures (in °C) (Figure 24.2)
3 Annual precipitation in New York State in the form of a moisture/water availability index (in mm) (Figure 24.3).

The temperature and precipitation values are averaged over a 50 years period (1950–2000) and were extracted from the WorldClim database (www.worldclim.org).

In addition to the maps (Figures 24.1, 24.2, and 24.3) all datasets are available to you in electronic format (GIS shapefiles for the freeware application ArcExplorer9.2® and MS Excel) on this book's website.

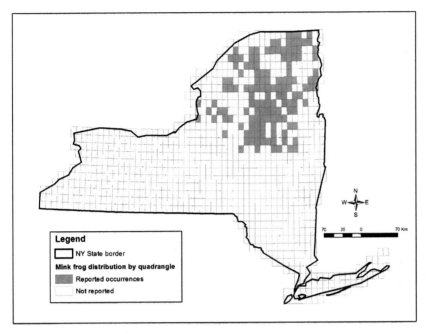

Fig. 24.1 Current range of the mink frog in New York State.
Source: New York State Amphibian and Reptile Atlas.

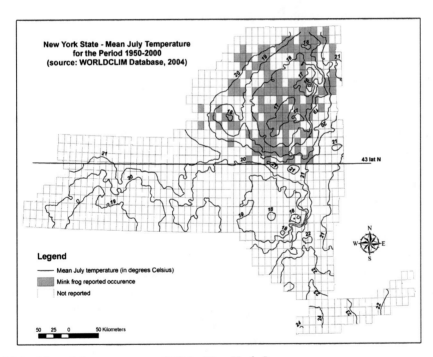

Fig. 24.2 Mean July temperature (°C) in New York State.

Note that the climate maps (Figures 24.2 and 24.3) are smoothed plots that look like topographical maps and are read in a similar fashion. The lines are contours referred to as "isoclines." The mink frog occurrence map (Figure 24.1) has been overlapped with the temperature and the precipitation contours and is based on the United States Geological Survey quadrangles (1:24 000 scale). Each of these

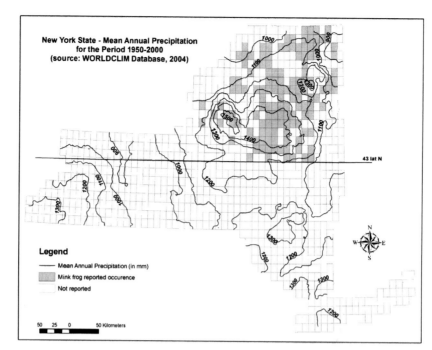

Fig. 24.3 Mean annual precipitation (mm) in New York State.

quadrangles covers an area bounded by two lines of latitude and two lines of longitude spaced 7.5 minutes apart. Given the New York State latitude, approximately 116 square kilometers are contained within one quadrangle. The corresponding MS Excel electronic dataset contains values of temperature and precipitation extracted for the center of each quadrangle. Separate sets of values for all of New York State and for just the quadrangles where the mink frog occurs are provided. GIS "shapefiles" that can be imported in ArcExplorer9.2® contain points that have as attributes the same values as the MS Excel files; there are two separate point shapefiles for: (i) all the New York State quadrangles (nyst.shp) and (ii) the quadrangles with only the mink frog occurrences (minkfrog.shp). Two additional polygon shapefiles of the New York State 7.5 minutes quadrangles (ny_quadindex.shp) and counties (nystate.shp) are also provided for visualization. All GIS files are archived; you need to unzip the files before using them (help is provided in the Making Spatially Explicit Predictions with ArcExplorer section below).

Last, you need information on likely climate change for the region. Based on projections for the next century made by the Intergovernmental Panel on Climate Change and results from the United Kingdom Hadley Centre's climate model (HadCM2), by 2100 temperatures in New York could increase by about 2–3 °C (or 4–5 °F). Precipitation is projected to increase by 10–20%.

With these data you can now determine:

1 where New York State currently lies within the general "climate space" of ambient temperature and rainfall
2 what subset of that climate space the mink frog currently occupies
3 given likely changes in New York's climate, whether the mink frog will be left with any habitat within its current climatic tolerance range.

There are several approaches to the problem and you might consider alternative ways to combine the different datasets.

Making General Predictions in a Spreadsheet Environment

Here are some suggestions on how to tackle the problem simply by using the MS Excel® datasets:

1 Plot where New York currently lies in "climate space" using the records for the mean July temperature and annual precipitation in the MS Excel file. Produce a bivariate plot (*x* versus *y* "scatter plot") in MS Excel (or some other graphing package). Draw a line around the extremes of this scatter of points – this is New York's current "climate space" with respect to these two climate variables.

2 Now plot where in this climate space the mink frog currently occurs and draw a line around that swarm of points. This represents the climatic conditions that currently constrain the species' range in the State.

3 Now project the change in temperature and rainfall for each grid point in the New York State climate space assuming uniform changes in climate at all points based on the midpoints of the projections described above. This is easily done by creating two new columns in your spreadsheet and manipulate the existing columns as appropriate (increase temperature by about 2–3 °C, i.e. add to current values for each quadrant 2.5, and increase precipitation by 10–20%, i.e. multiply current values for each quadrant by 1.15).

4 Draw a line around all the points that represent New York State's future climate space.

5 Last, determine what fraction of the current mink frog occurrences remain within the future climate space.

Making Spatially Explicit Predictions with ArcExplorer

Next, you can use all the electronic data and GIS software to locate the actual areas that would be likely to support mink frog populations in the future. For this task you can use the GIS software ArcExplorer9.2® (Environmental Systems Research Institute, ESRI) available for free download at the following URL address: http://www.esri.com/software/arcexplorer/download9.html. An ArcExplorer9.2® user manual can also be downloaded from this book's website.

Here are some hints on how to handle and analyze the data:

1 First download the data from this book's website. Create a new folder on your computer and name it "minkfrog." "Unzip" the data files provided into your folder using any archiving software (if you don't have one, you can download the WinRAR archiver free of charge from the following address: http://www.rarlab.com/download.htm). Once you unzipped the files, upload the shapefiles in ArcExplorer9.2® and visualize them. You do this by clicking the Add Files button ⬚ on the Menu bar and navigating to your folder; add more files at a time holding the Ctrl button and clicking on the files. We need the files named "minkfrog.shp" and "nyst.shp," which contain the climate information (Figure 24.4).

Remember that the mink frog points symbolize the quadrangles with "reported" occurrences only. This means that mink frog populations might exist elsewhere in northern New York State, but have not yet been detected. Once you have uploaded the files, visualize the datasets by checking the box located in the upper left corner of each file in the legend, next to the file name. To modify colors, dot sizes, and so on, double-click on an item in the legend and a Properties dialog will appear

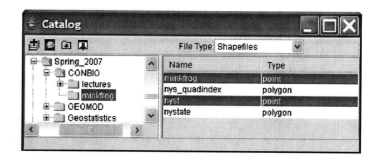

Fig. 24.4

(e.g. increase the dot size for the minkfrog.shp dataset from 6 to 12 to visualize the mink frog occurrences). Now, your map should look like the one in Figure 24.5.

The *nyst.shp* file contains 4 columns that are of interest: *July_Temp*, *Precipit*, *Temp_pred* and *Pp_pred*, which represent the *current mean July temperature*, *current annual precipitation*, *predicted July temperature*, and *predicted annual precipitation*, respectively. The last two columns are left blank for you to fill in (see point 3 below for more information on how to perform this operation).

2 Compute simple statistics (range, mean, standard deviation, minimum and maximum values, standard error of the mean, 95% confidence interval) for each climate parameter composing the current mink frog and the New York State climate spaces. This will help you understand how climate limits mink frog distribution and which of the two climate parameters is more important. You can do this operation in MS Excel, any other statistical package, or in ArcExplorer9.2® using the Query Builder button ⬛ from the Menu bar. Make sure that the point file you want to compute the summary statistics for is selected (e.g. in the image below, the first item in the legend, "minkfrog", is selected) before you start the Query Builder. Click on Statistics and then select the data you want (i.e. July temperature or annual precipitation; see Figure 24.6).

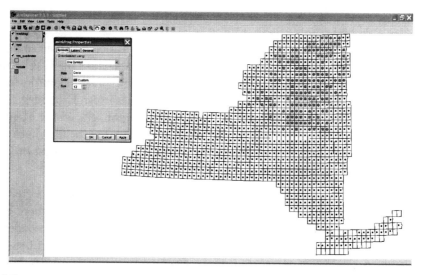

Fig. 24.5

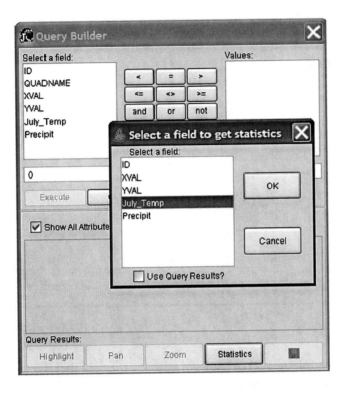

Fig. 24.6

The output contains the range, mean, and standard deviation of each climate parameter. It is up to you to derive the other simple statistics elements. Compare the mink frog requirements with the New York State climate space values. Which climate parameter is most likely to limit the mink frog distribution in New York?

3 Fill in the two blank columns in the New York dataset (nyst.shp) with the predicted data. You can do that easily by opening the database (dBase) file associated with the point file (nyst.dbf) in MS Excel and perform the operations as in a normal spreadsheet file. In MS Excel go to the Open File dialog, navigate to your folder and select only the dBase files to display (Figure 24.7).

Perform the same addition and multiplication operations as suggested when working with normal MS Excel files (Step 2 in Making General Predictions in a Spreadsheet Environment, above). Just make sure that you fill in the blank columns only and do not erase any columns or modify the columns names. After you filled in the appropriate predicted July temperature and annual precipitation, close the file and click Yes for everything MS Excel asks (save it with the same name, replace the existing file, and keep the dBase format even if MS Excel warns you of some incompatibility).

Close and reopen ArcExplorer in order for the changes in the nyst.dbf file to take effect. Add the shapefiles following the instructions from Step 1.

4 Now, based on your knowledge of mink frog climatic requirements and your judgment of the most limiting factor(s), you can use the Query Builder button to create selections of points (hence quadrangles of likely occurrence) that are likely to sustain mink frog populations. First, establish a series of thresholds for the mean July temperature and annual precipitations that in your opinion describe "optimum" mink frog conditions (i.e. mean values, median, mode or any other

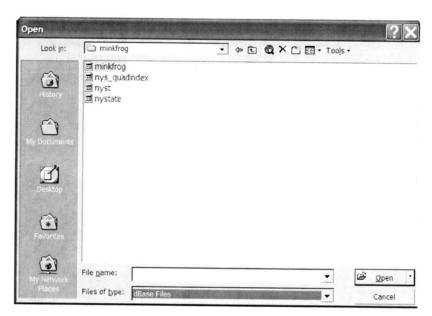

Fig. 24.7

value you consider relevant). You can also think of values that are either very restrictive (i.e. the lower bound of the 95% confidence intervals, the minimum temperature) or permissive (i.e. the upper bound of the 95% confidence intervals, the maximum values) for mink frog distribution. You will use these values as thresholds to investigate multiple predictive scenarios, since you are not certain which climate parameter and which values are critical for determining mink frog distribution patterns. For example, you might consider that the July temperature is the only important variable in predicting occurrence and that the 18.95 °C value represents a temperature threshold past which the mink frogs are less likely to persist. Hence, you will want to select and identify all possible future occurrences that have a predicted mean July temperature of less than 18.95 °C. Or you can consider the July temperature and annual precipitation simultaneously and perform similar selections (see below for details). Now you are ready to create predictive scenarios of mink frog distribution change as driven by the global warming.

5 To continue the case we started above (which is arbitrary and for illustrative purposes), you can use only one climate parameter (e.g. Pred_temp < 18.95 °C etc.) or both factors with the Boolean operators (e.g. Pred_temp < 18.95 °C AND Pred_pp > 1108.0 mm etc.). The first example can be defined as "select all points that have a predicted mean July temperature less than 18.95 °C", while the second denotes "selects all points that simultaneously have a predicted mean July temperature less than 18.95 °C and a mean annual precipitation greater than 1108.0 mm". In the Query Builder dialog, select the desired parameter(s) and the thresholds that you chose at step 4, and click Execute (Figure 24.8).

A map like the one below (Figure 24.9) should result. The light-colored points are the ones that were selected based on your Query and the number of points is given in the Query Builder dialog (i.e. Query results: 231 selected, for the first example above).

You can first create selections of points based on the *current* New York State climate. These points should show quadrangles that are likely to sustain mink frog populations at the present day. The point selection procedure remains the same, but instead of the *predicted*, you should use the *current* New York State climate values.

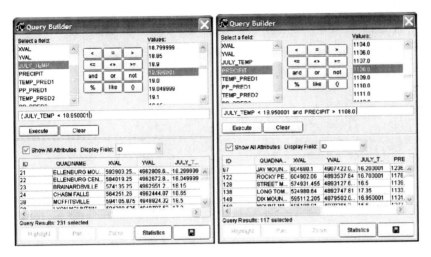

Fig. 24.8

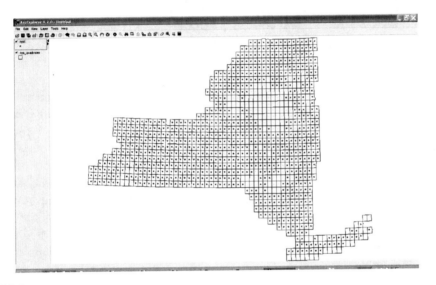

Fig. 24.9

Compare your selection of mink frog likely distribution with Figure 24.1 and comment on the disparities between the two maps.

Repeat the procedure for the *predicted* New York State climate using the same selection values you used before. Report the number of points (quadrangles) that were selected each time along with the values you used as thresholds. Also, you need to save each map you created (using the selections) as images, to compare distributions (see Step 6 below for more details). Note that there are two important issues to keep in mind when reporting the number and locations of your selected points:

(i) the mink frog is not likely to reach areas located south of its current location, since it has poor dispersal abilities and there are anthropogenic barriers (e.g. major highways) and other factors that we did not account for that would impede its dispersal;

(ii) the current locations are only "reported occurrences", so the mink frog could actually inhabit other quadrangles in northern New York State.

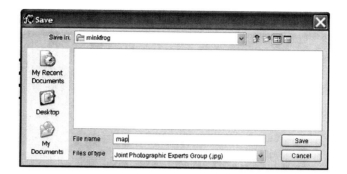

Fig. 24.10

6 You can repeat the entire process (starting from step 3) and redo the calculations in the nyst.dbf file to create a time series of predictions. Instead of adding 2.5 °C to the current temperature and multiplying the precipitation by 1.15 (representing predictions over 100 years), you can use intermediate values (e.g., adding 1.25 °C to the current temperature and multiplying the precipitation by 1.075) to give you an idea of what the mink frog distribution is likely to be over the next 50 years.

7 The final step in this analysis is to compare the number and location of possible NY State quadrangles left available for the mink frog based on the different temperature and precipitation thresholds for mink frogs that you used in your selections and for different time steps (if you chose to undertake Step 6). Now you are able to predict dynamic mink frog distribution scenarios given the potential effects of global climate change. This is most easily done by comparing the number of selected points given the thresholds you used and by visualizing the associated maps. To save and store the selections you created as maps (figures you can insert in your report), simply click the Copy Image to File button 🖼 on the *Menu* bar right after you performed a point selection. In the Save window (shown in Figure 24.10), navigate to your "minkfrog" folder and assign a relevant name according to the selection it represents. The only format option for saving is Joint Photographic Experts Group, also known as JPEG. Click Save; now you are able to import the image into MS Word and provide an explicit spatial representation of potential mink frog distribution given the global climate change scenarios.

Expected Products

- For MS Excel only users, graph(s) indicating the potential shifts in climate in relation to mink frog occurrence. For GIS users, maps showing future locations that would be available for the mink frog.
- A summary (in written or verbal form) of the methods you used to represent the mink frog redistribution and the rationale behind choosing the July temperature and precipitation values you used as thresholds for your selections
- Responses in a form indicated by your instructor to the Discussion questions below.

Discussion

1 How would you use your predictions of future mink frog locations for conservation purposes?
2 What other information would you need to improve your predictions?
3 Think of other species in your region that might have specific climate requirements or restricted distributions. Are they in peril of extinction due to global climate change?
4 This analysis assumes that mink frogs cannot adapt to the projected changes in magnitude and, in particular, rate of climate change. Do you think this is a reasonable assumption?

Making It Happen

Now that you are acquainted with basic bioclimatic modeling, it is your turn to think about how these methods may be applied to other organisms elsewhere. There are two issues to consider: first, what other areas are most at risk for climate change (e.g. arctic or tropical high-elevation areas) and second, what organisms are more vulnerable to climate change (i.e. species that have low dispersal abilities, that have strict habitat requirements, are more sensitive due to their behavioral or physiological traits, etc.). You can take this type of analysis further and use free software and climate data we recommend below.

Further Resources

More information on "climate space" / "bioclimatic envelope" analysis and freeware software can be found at www.diva-gis.org. This website also provides a large amount of global-scale free and downloadable data (climate, boundaries, land cover, altitude, satellite imagery etc.). You can take it from there and perform a multitude of climate-related analysis for plant and animal species of interest.

CHAPTER 25

Climate Envelope Modeling: Inferring the Ranges of Species to Facilitate Biological Exploration, Conservation Planning, and Threat Analysis

Robert Hijmans, Andy Jarvis, and Luigi Guarino

You would be rather startled to find a polar bear paddling in the Amazon or a palm tree swaying in the balmy breezes of Siberia. That's because most species have quite predictable distributions around the globe. Even species with really wide distributions such as ospreys or bracken ferns are not found in every ecosystem.

Species distributions are shaped by many factors: physical features of the environment such as climate and soils, the distribution of other species, and past and current geographic barriers. However, climate is usually reckoned to have a fundamental effect, and is increasingly being used to probe and predict the distributions of species. A technique called **climate envelope modeling** uses points where a species is known to occur to infer the climatic adaptation of the species, and then to identify the degree to which the climate in other areas is suitable for the species. See Graham et al. (2004) for more information.

This technique is of great interest to conservation because occurrence data from e.g. natural history collection databases are often sparse and do not fully describe the geographic distribution of a species. Climate envelope modeling fills in the gaps in the collections by providing an indication of the likelihood that a species might be found in unexplored areas. Specifically, this can be useful for:

- biological exploration, targeting sites which it might be worthwhile to visit to find a given species
- conservation planning, providing information as to the species likely present in a given area
- threat analysis, identifying the factors that might cause changes in the distribution of a species (for example, climate change).

For this exercise, 26 wild species of the genus *Arachis* (the relatives of *Arachis hypogaea*, the cultivated peanut), are used as an example. This exercise is partly based on Jarvis et al. (2003), which is useful to consult as you work through exercise.

Objectives

- To learn the basics of climate envelope modeling
- To examine the climatic adaptation of a species
- To predict the entire geographic range of a species.

Procedures

If you have not done so already, download and install on your computer the free GIS program called DIVA-GIS from www.diva-gis.org. It is assumed that you have completed the DIVA-GIS exercise (on wild potatoes) in chapter 23 of this book, which also means that you have probably read the first chapter of the DIVA-GIS manual. For the present exercise, it would be useful to read chapter 7. For this exercise you need to download climate data from the DIVA-GIS website (http://diva-gis.org/climate.htm); select the data that have 10 minutes spatial resolution and follow directions given on the web site.

Data Preparation

1 Make a new folder, and unzip the data for this exercise into it.
2 Make a shapefile called "peanuts.shp" of the points in "arachis_accessions.dbf" and save it in the new folder. This file has 397 localities representing collecting sites.
3 Add the shapefile to the map; also add the "pt_countries" shapefile (download from the book's website). See Figure 25.1.

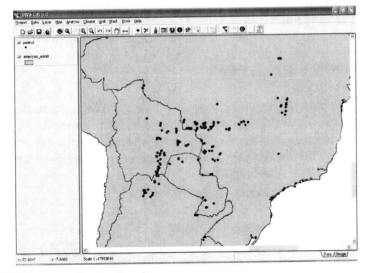

Fig. 25.1 Wild peanut distribution data.

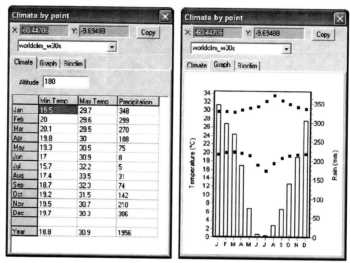

Fig. 25.2 Visualizing the climate data.

4 If you do not have them already, now download the global climate files at 10 minute resolution for current conditions. Check if you have climate data files by clicking on Climate > Point and then clicking on any part of the map (where there is land). If you cannot get windows like the two shown below (and get the error message "no climate databases found"), you should download the climate data from http://www.diva-gis.org *OR* you need to change the folder where DIVA-GIS looks for the climate data (under Tools > Options > Climate). See Figure 25.2.

Distribution Modeling

Frequency distributions

Now we are going to explore the climate data associated with the peanut accessions that we have imported.

1 Make the "peanuts" layer the active layer.
2 Use Modeling > Bioclim and Domain.
3 Go to the second tab *(Frequency)* and press Apply. See Figure 25.3.
4 Note that there are 161 non-duplicate observations. Points that are at exactly the same location are included only once. Points that are at a different location but in the same cell of the climate grid are also included once, in this case (this is an option that can be changed on the first tab; if you do so you will note that there are 224 unique observations).
5 Explore this tab by changing the climate variables. This is helpful to get an idea about the general characteristics of the distribution of these points in environmental space (as opposed to the distribution in geographical space which you can see on the map). It may also reveal environmental outliers, perhaps caused by incorrect coordinates.
6 If you click on a point in the graph different things can happen, depending on the state of checkboxes above the graph. Check this out.

Environmental Envelope

1 Go to the Envelope tab. Here you can look at the distribution of the points in two dimensions of "environmental space."

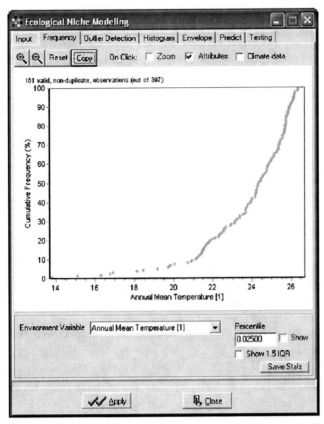

Fig. 25.3 A cumulative frequency distribution for the mean temperature data.

2 Display different "environmental envelopes," using the Percentile button. Consult the textbox below entitled Climate Envelope Modeling for further background. For example use percentiles 0, 0.05, 0.1, and 0.25. The points inside the multidimensional (i.e. all climatic variables) envelope at the percentile cut off specified are colored green; the other ones are colored red. Thus, red points within the envelope must be outside of it in another dimension (with regards to another variable on the horizontal or vertical axis).

- The *percentiles* are calculated for each variable individually, and then combined. The *percentage* of the points that are inside a multidimensional envelope at a certain percentile will be different for each data set.

- The points that are inside the (two-dimensional) envelope on the graph are now colored yellow on the map. We have thus established a link between the distributions in ecological and geographic space (Figure 25.4).

Modeled Distributions

1 Go to the Predict tab.
2 First, enter the area for which you want to make a prediction. Use these values: MinX = −71; MaxX = −39; MinY = −33; MaxY = −4.
3 Select the *Bioclim* output, choose an output filename, and press Apply. The result should be as in Figure 25.5. Note that this result suggests that there is a large area in south central Brazil where the climate appears to be suitable for peanuts but where no peanuts have been collected (or even occur?). See Figure 25.5.

Climate Envelope Modeling

In climate envelope modeling (also referred to as species distribution modeling or ecological niche modeling) a set of points where a species was observed is used to infer its environmental requirements. For example, when a species only occurs in cold and dry areas then the reasonable assumption is made that the species can only survive under such conditions. These inferred environmental requirements can then be used to predict the likelihood of the species being present (or able to survive) at any place or time period for which environmental data is available. See Graham et al. (2004) for more information.

Many different statistical methods have been applied to this problem, including distance metrics, percentile distributions, logistic regression, principal components analysis, and machine learning approaches such as neural networks. Many of the early methods use "presence-only" data, but more recently developed methods tend to use "presence/absence" data. Sometimes the absence data used is simply a random sample from throughout the study area. Elith et al. (2006) compared many methods and showed that machine learning and advanced regression methods generally outperform other methods.

In this exercise we use the BIOCLIM model, because it is implemented in DIVA-GIS and easy to understand and run. DIVA-GIS also implements the DOMAIN model (see the manual about more information about these models). Both are presence-only models that may not perform very well. Maxent is an easy-to-run model that is thought to be much superior (see Further Resources).

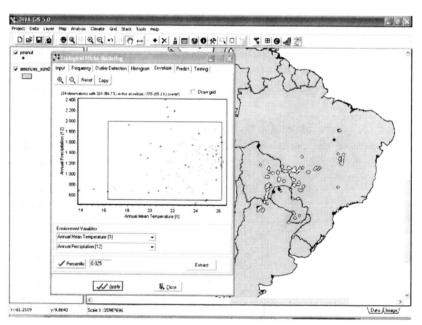

Fig. 25.4 Establishing a link between the distributions in ecological and geographic space.

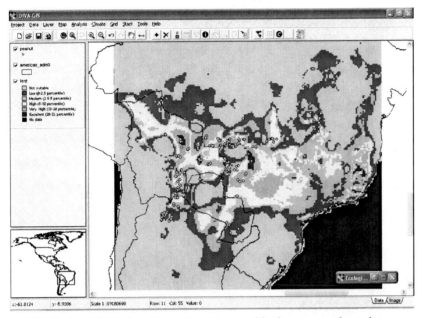

Fig. 25.5 Predicting where the climate is to be suitable for peanuts but where no peanuts have been collected.

4 Now make distribution models with different output types (use Bioclim True/False and Bioclim Most Limiting Factor Note that they are all representations of the same result, and that the result is dependent on the percentile cut-off you choose (explore this cut-off in the graph of the Envelope tab). The Most Limiting Factor outputs the variable with the lowest score in each grid cell for which there is a prediction.

Single Species Models

1 So far, we have treated the data as representatives of a single taxon, the genus *Arachis*. Now let's go to the species level. Go to the Input tab, select Many Classes and then select the Taxon field (see Figure 25.6). Go over the list of taxa to assure that there are no errors. A typical problem is that the same species occurs more than once because of spelling mistakes.
2 Now go to the Frequency tab and select one or two classes (or species, in our case). In the example below we compare the annual rainfall for two species. *A. kuhlmannii* is clearly distributed in drier areas than *A. stenosperma*. Note that there is one huge outlier in the rainfall distribution of *S. kuhlmanii*. This could be a problem.

Click on that outlier on the graph, to find out what record it represents, what its associated climate data are, and where it is located on the map (use the checkboxes above the graph). For example, first select the species, using Layer > Select Records so that the records for that species are colored yellow on the map. Then click on the outlier on the 'frequency' graph, and see to which point on the map it corresponds. It is clearly a geographical outlier as well. No members of the species (or any other *Arachis* species for that matter) have been observed in this area. In this case it would be prudent to check the coordinates against the locality description. However, that information is not provided here so there is not much we can do to assess whether this record is valid or not. Very much a problem! It is up to your "expert opinion" whether to leave this record in or take it out. Let's leave it in for this exercise (Figure 25.7).

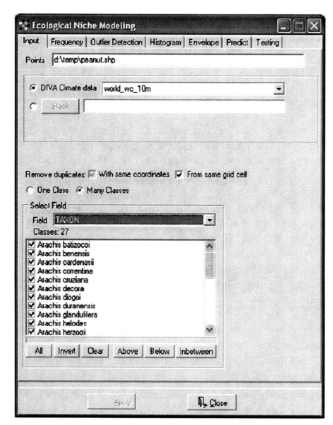

Fig. 25.6 Selecting a particular taxon of wild peanut for modeling.

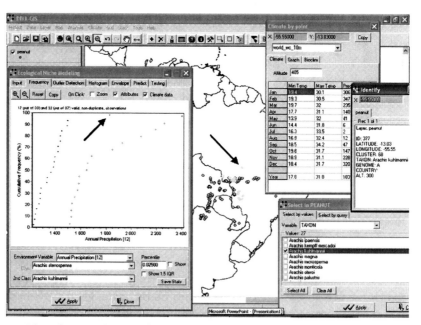

Fig. 25.7 Identifying outliers in climate space.

3 Go to the Predict tab and make modeled predictions (using Bioclim) for the same area as before (set the coordinates again as before, or make the previously modeled distribution the active layer and press the Read from Layer button to get the coordinates). In this case, however, use the Batch option. This will create a

prediction for each species. Save the output file in a new folder. Note that the output file is a **stack**. A stack is a collection of one or more gridfiles with the same origin, extent, and resolution.

4 Open some of the gridfiles that you have just made and compare the modeled range map with the observed points. There are a number of ways to look at the points for one species at a time. One option is to select the records for the species in question. The default color for the selected points is yellow, which, in this case, is also a color on the grid. You can change the color of the selected records under Tools > Options. You can also save the selected records to a new shapefile. Try it.

However, the easiest way to show a single species is by using Layer > Filter. Make the peanuts shapefile the active layer, then select the field Taxon and the value *A. stenosperma*. *A. stenosperma* has a disjunct distribution (a number of populations along the coast and another group more inland). There is one record right in the middle of the two distributions. This record again warrants checking. If it is correct, it would make an interesting specimen for study. A disjunct distribution can be difficult to model. It may very well be that the groups are in fact different ecotypes with a quite distinct climatic adaptation. In this case, the predicted range has all the inland localities on its margins, and shows a large area of apparent suitability from where there are no records of this or any other *Arachis* species. Also note some small isolated areas far away from the main area of the observed records and predicted range. These areas are probably irrelevant because they are very small, and far away from any locality where the species has been observed (Figure 25.8).

Modeled versus observed diversity

1 Add all the gridfiles (Stack > Calculate > Sum). Use the Sum as Present/Absent and > 0 option. Only very few areas are predicted to have more than one species (Figure 25.9). Are all areas included that were on the prediction for the genus as a whole? With other model settings, the results can change quite a bit. For example,

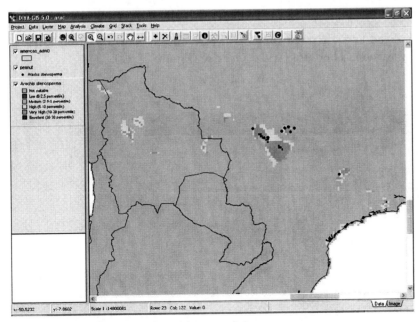

Fig. 25.8 Observed and predicted distribution of *A. stenosperma*.

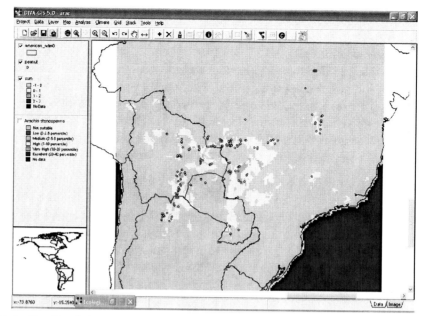

Fig. 25.9 Predicted diversity pattern.

repeat the modeling and summation using only bioclimatic variables 1, 5, 6, 12, 15. (Uncheck the others under Select Variables on the Predict tab.)

2 Aggregate (Grid > Aggregate) the stack to a 1 degree resolution (6x) using the MAX option and sum the result again. Aggregation combines smaller cells into larger cells. It will frequently be done using the Average option so that the larger cell has the mean value of the original smaller cells. Here we use Max because if is species is present in only one small sub-cell of the large cell we can say it is present in the larger cell (Figures 25.10 and 25.11).

3 Make a map with observed species richness at the same resolution as the previous map. Use Grid > Point to Grid > Richness (use Define Grid, and "Use parameters from another grid" to assure having exactly the same grids).

Compare modeled and observed richness before and after aggregation. First use Analysis > Regression. Then use Grid > Overlay (Figure 25.12).

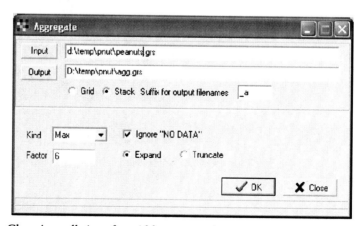

Fig. 25.10 Changing cell size of a grid by aggregation.

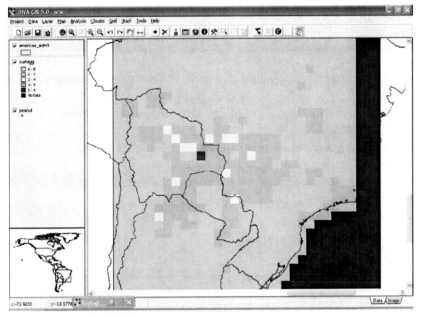

Fig. 25.11 Aggregated grid data.

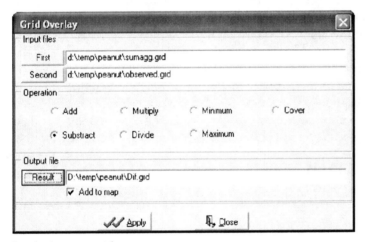

Fig. 25.12 Overlaying two grids.

Expected Products

- Description of the climatic adaptation in "ecological space" for the genus as a whole and for two contrasting species
- Map of the predicted distribution for one species of the genus
- Map of the observed and predicted species richness of the genus
- Map identifying new areas where further exploration of a species might be successful
- Responses in a form indicated by your instructor to the Discussion questions below.

Discussion

1 What other factors apart from climate might be important in defining a species range?
2 How many collection points do you think are needed for climate envelope modeling to be accurate?
3 Can you identify the potential sources of error that might cause the results to be less accurate?
4 How would you analyze the climatic envelope of species with disjunct distributions, or with known ecotypes?

Making It Happen

You can refine this exercise by using different climate envelope models and combining the results to increase the confidence in your predicted species range. Some of these are available in DIVA-GIS (DOMAIN, for example), whilst others require installation of other programs (MaxEnt or GARP, for example) although DIVA-GIS can be used to prepare the input files for your analysis. You may also want to experiment with methods for validating your results. This can be done by split-sampling of the input collection point data, and using statistical analyses such as Kappa or AUC plots. This is available in DIVA-GIS in Modeling > Evaluation. You could also combine the predicted range from the climate envelope model with other datasets such as land cover to limit the range depending on the habitat preference of the species under study. Finally, you can bring all the analyses together to identify conservation priorities through overlaying a protected areas dataset and examining the current effectiveness of the reserve network in conserving the species found in the genus. Species that do not fall in protected areas are then priorities for conservation and the predicted ranges can be used to identify the regions where a reserve would potentially conserve the greatest biological richness.

Another interesting application would be to look at climate change, but that is discussed in exercise 24 of this book.

Further Resources

For those interested in exploring these issues further, the following resources are recommended:

- Maxent species distribution model: www.cs.princeton.edu/~schapire/maxent/
- More data on climate: www.worldclim.org
- More data on land cover: www.diva-gis.org/Data.htm or https://zulu.ssc.nasa.gov/mrsid/
- More data on protected areas: www.unep-wcmc.org/wdpa/.

PART 6

Policy and Organizations

Population, Consumption, or Governance: Which Drives Species Imperilment Most in Africa and Europe?

James P. Gibbs

Six billion, six hundred and seven million, four hundred and ninety thousand (6,607,490,000) people consuming three hundred and twenty-six thousand, two hundred and ninety-six quadrillion (326,296,000,000,000,000,000) joules of energy and using three trillion, six hundred and eighty-seven billion, ninety-four million (3,687,094,000,000,000) cubic meters of water per year. Numbers like these are baffling. On the one hand, they seem to quantify precisely the extent of the world's environmental problems. On the other hand, they totally defy any attempt to really grasp them, to get your mind around them. Few people have an intrinsic feeling for what a "million" is, let alone more than a million million cubic meters of water or joules of energy. The one thing that we can say about the numbers that define the human overpopulation problem and our overuse of the world's resources is that they are indeed over the top.

To put it simply, there are just far too many of us using too many resources. Not every one recognizes that overpopulation and overconsumption are two sides of the same coin. Some people from rich nations with modest populations point to the world's poor, rapidly growing nations as the primary source of global environmental problems. People from poor nations tend to blame excessive consumption by wealthy nations. It is futile trying to partition the blame; we all need to work on both sides of the problem.

So what are the causes of biodiversity loss? We often focus on the **proximate causes**: unsustainable resource use, habitat fragmentation and loss, invasive species, pollution, climate change, and chains of extinctions. Dealing with these threats is the daily pre-occupation of conservationists. But conservation biology has often been criticized for not focusing on the **ultimate causes** and, like Nero, "fiddling while Rome burns." These ultimate causes have to do principally with us – humans – and our changing numbers, patterns of resource consumption, and ways we manage our affairs (i.e., governance) (McKee et al. 2004, Smith et al. 2003). The purpose of this exercise is to cut right to the point and examine these ultimate factors to ask: What drives most species imperilment... human density, economic development, or governance? We ask this question by combining information from two regions of the world – Africa and Europe – that contrast strikingly in terms of their degree of wealth, population pressures, and patterns of governance.

Objectives

- To examine how essential aspects of human society affect the status of a high-profile biodiversity indicator group
- To think broadly about the changes in human society required to retain significant amounts of biodiversity in the environment.

Procedures

The Nature of the Data

We have assembled information by country on the status of a biodiversity indicator group, three basic aspects of human society, and an ancillary biodiversity control variable (Table 26.1). The full data set is available at the book's website. Information was assembled for countries in Europe and Africa that were more than 10 000 km² in area and for which a complete set of relevant data were available. We focus on three main attributes of human society – population density, gross domestic product per capita, and an index of integrity of governance – as ultimate drivers of biodiversity change. For a response variable, birds were chosen as an indicator group because they are perhaps the only taxon that is sufficiently well inventoried around the globe to permit this kind of analysis.

Sources of Data

Actual sources of data are outlined in the Further Resources section below. For **birds,** we obtained from the 2000 IUCN "Red List" of threatened species both the number of threatened species (status of CR – critical, EN – endangered, or VU – vulnerable) and the total number of extant species. As an overall response variable total threatened species was adjusted by country area, that is, number of threatened species per 10 000 km² to control for larger countries having more threatened species simply because they covered more area. We have included also total number of extant birds in each country as a **"control" variable.** Countries that were considered host widely varying levels of background biological diversity across two continents. Including the total bird species for each country essentially can control for the possibility that lower levels of imperilment could simply be due to lower levels of diversity (and vice versa).

For human societies, we assembled data on **gross domestic product** (nominal) per capita, which is the value of all final goods and services produced within a nation in a given year, divided by the average population for the same year. This is a widely used measure of typical incomes and overall economic development of a country. These data were obtained from the International Monetary Fund, World Economic Outlook Database for September 2006. An index of the **quality of governance**

259

Ch. 26
Population,
Consumption,
Governance

Population, Consumption, or Governance

Table 26.1 Data set examining drivers of species imperilment in Africa and Europe, including density of threatened bird species (response variable), human density, economic development (GDP per capita) and governance (CPI score), with total bird species richness as a control for background biodiversity levels. (Note: these data are also available for download at the book's website.)

Country	Threatened Species per 10,000 km²	Human population density (people/km²)	CPI Score	GDP (nominal) per capita	Total bird species recorded
Africa					
Benin	0.09	72	2.5	592	112
Botswana	0.05	3	5.6	6439	184
Burkina Faso	0.04	52	3.2	430	138
Burundi	0.72	301	2.4	107	145
Cameroon	0.08	38	2.3	952	165
Central African Republic	0.02	7	2.4	335	168
Chad	0.02	8	2.0	654	141
Congo (Kinshasa)	0.00	28	2.2	615	130
Cote d'Ivoire	0.19	56	2.1	900	252
Ethiopia	0.06	69	2.4	153	262
Gabon	0.07	5	3.0	6397	156
Ghana	0.13	96	3.3	512	206
Guinea	0.12	40	1.9	355	109
Kenya	0.17	64	2.2	560	344
Libya	0.01	3	2.7	6696	76
Malawi	0.42	115	2.7	161	219
Mali	0.02	10	2.8	432	191
Mauritania	0.01	3	3.1	663	172
Morocco	0.20	76	3.2	1713	206
Niger	0.01	10	2.3	274	125
Nigeria	0.04	146	2.2	678	286
Rwanda	1.52	376	2.5	242	200
Senegal	0.10	64	3.3	738	175
Sierra Leone	0.56	86	2.2	223	172
South Africa	0.08	36	4.6	5106	304
Sudan	0.01	17	2.0	820	280
Uganda	0.17	126	2.7	303	243
Zambia	0.07	15	2.6	627	252
Zimbabwe	0.10	32	2.4	383	229
Africa average	0.17	67.4	2.7	1312.4	194.6
Europe					
Armenia	1.34	100	2.9	1140	236
Austria	0.36	98	8.6	37117	230
Azerbaijan	0.81	94	2.4	1493	229
Belarus	0.14	47	2.1	3031	194
Belgium	0.66	340	7.3	35712	191
Bosnia and Herzegovina	0.59	89	2.9	2384	205
Bulgaria	0.90	66	4.0	3459	248
Croatia	0.71	79	3.4	8675	224

(*Continued*)

Table 26.1 (*Continued*)

Country	Threatened Species per 10,000 km²	Human population density (people/km²)	CPI Score	GDP (nominal) per capita	Total bird species recorded
Czech Republic	0.25	130	4.8	12152	205
Denmark	0.23	127	9.5	47984	196
Estonia	0.67	29	6.7	9727	204
Finland	0.09	15	9.6	37504	243
France	0.09	111	7.4	33918	283
Germany	0.14	231	8.0	33854	247
Greece	0.53	81	4.4	20327	255
Hungary	0.86	107	5.2	10814	208
Italy	0.17	193	4.9	30200	250
Latvia	0.46	35	4.7	6862	216
Lithuania	0.61	55	4.8	7446	201
Netherlands	0.96	399	8.7	38618	192
Norway	0.05	12	8.8	64193	241
Poland	0.13	123	3.7	7946	233
Portugal	0.87	116	6.6	17456	235
Romania	0.34	93	3.1	4539	257
Slovakia	0.82	111	4.7	8775	199
Slovenia	0.49	99	6.4	16986	201
Spain	0.14	80	6.8	27226	281
Sweden	0.04	20	9.2	39694	259
Switzerland	0.48	183	9.1	50532	199
Ukraine	0.12	77	2.8	1766	245
United Kingdom	0.08	250	8.6	37023	229
Europe average	0.46	119	5.9	21243	227.0
Overall average	*0.32*	*92.39*	*4.35*	*11610.22*	*211.30*

in each country was obtained from Transparency International (TI). TI is an international non-governmental organization addressing corruption and compiles annually surveys that ask businessmen and analysts, both in and outside the countries they are analyzing, their perceptions of how corrupt a country is. We use its Corruption Perception Index (CPI) for 2006. The index ranges from low values of transparency (indicating high corruption) to higher values indicating more transparency in internal affairs. Last, data on **human density** was obtained from for the year 2005 from the world database of the US Census Bureau.

Analysis

These data are most appropriately analyzed by means of a multiple linear regression. A multiple linear regression models the relationship of one variable y given the values of some other variables x. In this case, density of threatened bird species is the y (or response or dependent) variable and the others – human population density, GDP per capita, CPI, and bird species richness – are the x (or independent) variables. This approach assumes that variation in threatened birds will be a linear function of the other parameters, which is a reasonable starting assumption.

261

Population, Consumption, or Governance

Ch 26
Population,
Consumption,
Governance

Table 26.2. Multiple regression analysis of effects of human density, consumption levels, governance, and background bird diversity levels on variation in density of threatened birds (Table 26.1).

Parameter	Est value	St dev	t	Prob(>|t|)
Intercept	−0.11495	0.193	−0.595	0.554
Human density	0.00235	0.000	5.305	0.000
CPI score	0.08099	0.041	1.956	0.056
GDP per capita	−0.00002	0.000	−2.475	0.016
Bird species richness	0.00021	0.001	0.275	0.784
R^2(adj)	0.323			
F	8.051			
Prob(>F)	0.000			

P value

You will need access to some statistical package to calculate the estimates for this dataset. There are many options; we suggest using the free PopTools add-in to MS Excel® (www.cse.csiro.au/poptools/), but you or your instructor can likely identify many other options for performing least-squares multiple linear regression as either a stand-alone piece of software or as an internet-based application. If you do use PopTools you can find the easy-to-use regression function under PopTools > Extra Stats > Regression. Your results should look close to those in Table 26.2. Although we have provided you with the results of the analysis, an important aspect of this exercise is learning to do these analyses yourself, so we strongly encourage you to perform the multiple regression on your own.

Interpretation

Here are the main considerations when interpreting results of multiple regression.

- **Model "significance"** This is estimated by the F and Prob(>F) values. If the Prob(>F) value is <0.05, then you can conclude that a statistically significant amount of variation in threatened bird species density is explained by the variables in the model.
- **Model "fit"** This is approximated by the R^2(adj), which is the adjusted coefficient of determination, an approximate measure of the proportion of variability explained by the regression model. Values close to zero suggest a poor model, whereas values near one suggest a model that explains virtually all the variation.
- **Model coefficients** These indicate the extent to which individual variables drive variation in the response variable. The signs of these coefficients are important: a positive sign on a coefficient indicates that as the value of the independent variable increases, so does the response variable; a negative sign indicates that as the value of the independent variable increases, the response variable decreases. The "Prob(>|t|)" indicates which are significant (P < 0.05) predictors. And, quite usefully, the coefficients can be combined into an equation to predict the threatened bird species density on the basis of particular values of the independent variables, as follows:

$$-0.115 + 0.002 \times \text{Human density} + 0.081 \times \text{CPI Score} - 0.00002$$
$$\times \text{GDP} + 0.0002 \times \text{Bird species richness}.$$

Expected Products

- Evidence of original multiple regression analysis of the data Table 26.1 if so required by instructor
- Responses in a form indicated by your instructor to the Discussion questions below.

Discussion

1 So which continent has the bigger, current population "problem" (contrast average human densities between continents)? If overall human impact is a product of population density and consumption rate, what do you conclude?

2 Are human density, patterns of governance, and levels of economic development relevant to conservation biology, even based on this simple analysis at a coarse geographical scale? In other words, is the regression model statistically significant?

3 How good is the overall model "fit"? Approximately how much variation remains to be explained by other variables not included in the model?

4 Which independent variables are significant predictors of variation in status of birds? How do you deal with a variable like CPI score, which is "almost" significant according to the $P < 0.05$ convention employed in conservation biology and wildlife management?

5 Focus in particular on interpreting the relationship between GDP and bird species threat... is the coefficient negative or positive? Explain this relationship in simple terms.

6 Now for the main question... What would likely have a greater impact on the status of global bird life: halving human population density or doubling per capita GDP? (Hint: to answer this "plug" the average values for each parameter (see bottom of Table 26.1) into the overall regression equation and predict bird species status. Then double the average GDP and predict bird status again. By what factor has bird status changed? Now re-insert the average GDP value and then insert a population density value that is half of the average. By what factor has bird status changed?). From a general policy perspective where should the emphasis be placed: population control of poverty alleviation?

7 How do you reconcile increasing economic development to reduce bird species threat? Place your answer in the context of sustainable development. By extension, is one country's economic advance, which evidently improves the status of its bird life, independent of another's? Explain, with some concrete examples.

8 Are population growth and economic development related? In what way? Are these factors synergistic?

Making It Happen

Human overpopulation, consumption, and governance are the overriding issues of the day. Using this exercise as a springboard, consider how you might further involve yourself in these matters by addressing these additional, more general questions. How could your country improve its population policy? How about its overriding economic development policy? Which do you think has more of an impact on biodiversity, the developed countries that use resources unsustainably or the developing countries with high population growth rates? How could your actions help or hurt global biodiversity? How might your impact on the environment be different now, when you are a student, compared with when you lived at home or when you graduate? What do you think you can do to lessen your negative impact on the environment?

263

Population, Consumption, or Governance

Ch 26
Population,
Consumption,
Governance

Further Resources

Key resources used in this exercise that are worth consulting for further exploration are:

- International Monetary Fund, World Economic Outlook Database: http://www.imf.org/external/pubs/ft/weo/2006/02/data/index.aspx
- Transparency international's Corruption Perception Index 2006: http://www.transparency.org/policy_research/surveys_indices/cpi/2006
- US Census Bureau International Data Base (IDB): http://www.census.gov/ipc/www/idbnew.html
- IUCN Red List of threatened species: http://www.iucnredlist.org/

We also recommend McKee et al. (2004) and Smith et al. (2003) for more background on the issues raised in this exercise.

CHAPTER 27

Overconsumption: Who's Smarter... Students or their Professors?

James P. Gibbs

One philosophical tenet upon which virtually all agree is that one's actions should be consistent with one's beliefs. A painful dissonance can build if you say one thing but do another. Moreover, one's credibility suffers. Such was the case when it was reported that the home of Al Gore, a leading voice against global warming, uses some 191 000 kilowatt-hours of electricity per year when the typical household in the city where the Gores live uses some 15 600 kilowatt-hours per year (Associated Press, February 28, 2007). By extension, most people who work in the area of conservation biology and wildlife management presumably seek to live in a manner that does not harm the very creatures they have devoted their lives to conserving. And presumably the more education an individual conservationist has about the effects of human activities the more they might curtail destructive habits in which they engage.

Here we focus on one activity so emblematic of overconsumption of natural resources – excessive driving of automobiles. Driving has some clearly observed direct effects, such as flattened fauna, but also many other indirect effects that are cumulative and involve profound changes to ecosystems. More specifically, driving cars propels the construction and maintenance of roads, which together underpin many aspects of habitat fragmentation and loss. Moreover, light cars and trucks driven by individual citizens generate the bulk of CO_2 emissions from industrialized countries, which in turn disproportionately contribute to global climate change.

The short of it is that excessive driving of cars is bad for biodiversity no matter how you look at it. So how do people adapt their driving behavior based on their knowledge of biodiversity issues? One good indicator is how far they commute to work each day. The single decision of where to live in relation to workplace has huge ramifications for the impacts one has on the environment.

We have chosen as the focus for this exercise a small college located in a city in upstate New York. This college, like many others, is made up of a variety of individuals with different educational backgrounds. There are professors, presumably with advanced training in their field of specialty. There are also undergraduate students busily improving their knowledge base. There are graduate students specializing in a particular area. And there are administrators running the programmatic side of the organization. And then there are staff tending to the day-to-day operations. Neither of the latter groups typically has much formal training in environmental issues *per se*.

The particular college we have focused on – the College of Environmental Science and Forestry at State University of New York (SUNY-ESF) – is somewhat distinct in that all aspects of the college are focused on "the environment," be it environmental biology, forestry, chemistry, environmental studies or landscape architecture. All members of the college community are therefore immersed in environmental issues. Contrasting commuting behavior among them let us evaluate, acccording to the basic notions outlined above, the extent to which individuals' actions are in accordance with their presumed understanding of their actions' effects. More specifically, we test the notion that professors, all with PhDs and training in environmental science, and years of experience and exposure to the complexity of environmental issues, including the effects of transportation on biodiversity, would drive less than undergraduates, who, at first blush, would seem to know less about these issues. We also use graduate students, staff, and program administrators as comparison groups. Care to make any predictions about who drives most? Least?

Objectives

- To analyze patterns of commuting behavior among groups of environmentally oriented professionals to evaluate the proposition that increased amounts of formal education shape consumption behavior in positive ways.

Procedures

Base Data

We have summarized one-way commuting distances (in miles) for 35 undergraduates randomly polled on the campus "quad" during April 13, 2006 along with all 211 graduate students, all 104 professors, all 288 staff members and all 39 administrators at SUNY-ESF in 2006 (Table 27.1). You will need to fill in the remaining, blank cells in the table according to the directions in the text.

Your job is to summarize these estimates and examine whether they are statistically different from one another.

Calculating Confidence Intervals About the Estimated Means

Calculate the 95% confidence intervals about the estimates of mean commuting distances for each group. Generally speaking, if the confidence interval about the estimate for one group does not overlap that for another group, then the means are considered to be different. Confidence intervals for large-sized samples (n > 30, such as these) are calculated as:

$$95\% \ CI = M \pm (1.96 \times SE) = M - (1.96 \times SE) \ to \ M + (1.96 \times SE)$$

where M = Mean and SE is the standard error of the mean, which equals $SD/\sqrt{n}$.

Table 27.1 Average one-way commuting distances (in miles) at SUNY-ESF, a small, college focused on the environment and situated in an urban area in upstate New York.

	Under-graduates	Graduate students	Professors	Staff	Administrators
One-way commute (miles)					
Mean	2.5	8.8	8.7	12.7	14.4
SD	3.4	20.2	11.5	11.0	15.7
n	35	211	104	288	39
SE					
95% UCL					
95% LCL					
Extrapolation of environmental effects					
Annual CO_2 (pounds) 95% UCL					
Annual CO_2 (pounds) 95% LCL					
Annual road-killed vertebrates (individuals) 95% UCL					
Annual road-killed vertebrates (individuals) 95% LCL					

Extrapolating CO_2 Emissions

Next calculate the overall CO_2 emissions associated with commuting for each group for a given academic year (2 semesters). The US EPA estimates that 0.916 pounds of carbon dioxide are produced per mile driven (see Further Resources). Note that you will have to double the average one-way commute to obtain a daily round-trip estimate. Then multiply by 130 (13 weeks per semester × 2 semesters × 5 days per week) for a conservative number of daily commuting round trips per year associated with commuting. To make these predictions you only need to adjust the confidence intervals as appropriate.

Extrapolating Road-kill

Assuming that there are (i) 1.5 trillion miles driven each year in the United States, (ii) 1 million animals killed on roads each day, (iii) 52.2 weeks in a year and 7 days in a week (see Further Resources), how many road-kills are produced, on average, each academic year by each group? Here's a hint: based on the estimates above, there are some 0.00024 vertebrate animals killed per mile driven. As with CO_2 emissions, to make these predictions you just need to adjust the confidence intervals accordingly.

Expected Products

- Calculation of 95% confidence intervals for daily one-way commute, annual CO_2 emissions associated with commuting, and annual vertebrate road-kill
- Inference about difference among groups – undergraduates, graduates, professors, staff, and administrators – about individual annual environmental costs associated with commuting
- Responses to the issues raised in the discussion questions below in a format outlined by your instructor.

Discussion

1 So what's your answer to the question posed in the title of this exercise: Who's "environmentally" smarter...undergraduate students or their professors? How about the other groups sampled in this college community?
2 Why do you suppose commuting behavior varies among these groups? Does level of formal education matter? Does level of personal wealth matter more? It's dangerous to generalize, but assume that (i) the professors have much advanced learning (and degrees) about environmental matters and get paid relatively well, (ii) the administrators don't usually have advanced formal training in environmental matters but do get paid relatively well, (iii) staff don't usually have advanced formal training in environmental matters and get paid relatively poorly, (iv) graduate students are immersed in advanced training in environmental matters and get paid relatively poorly, and (v) undergraduate students are immersed in formal training in environmental matters and most are deeply in debt! What do you conclude "drives" commuting behavior most?
3 Are we as environmental professionals living in a manner consistent with our values and commitment to conserving wild species and the environment in which they live?

Making It Happen

The point of this exercise is to raise the question of whether we live in accordance with our beliefs about the importance of biodiversity and, if not, to ask why not. You may take umbrage with this oversimplification of such a complex issue. This exercise could be easily repeated and improved at your own locale to look at variation in commuting behavior, or other behaviors associated with heavy per capita consumption of natural resources, such as meat-eating (e.g., what percent of each group is vegetarian?), that characterize life in the developed world. The next exercise (28, on Conservation Values) can provide much guidance on designing effective surveys.

This exercise can make some students and professors uncomfortable because it can reveal hypocrisy and generate defensiveness. Others may find it irrelevant or intrusive ("Where you live is a personal matter!"). But the exercise's main intent is to raise the complex issues surrounding highly resource-intensive consumption behaviors and why humans make the choices that they do. If we do not come to grips with excessive resource consumption in industrialized societies then we will eventually lose the battle to conserve substantial amounts of biodiversity, no matter what happens with population growth in poorer, developing countries.

Further Resources

You might wish to pursue further some of the primary resources from which key data for this exercise were assembled:

- US Environmental Protection Agency Average Annual Emissions and Fuel Consumption for Passenger Cars and Light Trucks: http://www.epa.gov/otaq/consumer/f00013.htm
- US Federal Highway Administration Highway Statistics 2005: http://www.fhwa.dot.gov/policy/ohpi/hss/index.htm
- High Country News (estimate of annual roadkill in the US, February 7, 2007): http://www.hcn.org/
- State University of New York College of Environmental Science and Forestry: www.esf.edu.

CHAPTER 28

Conservation Values: Assessing Public Attitudes

Margret C. Domroese and
Eleanor J. Sterling

Consider the wolf, that awe-inspiring inhabitant of forest, mountain, and tundra whose calls on a moonlit night can evoke wilderness like no other sound on earth. This aura has made the wolf a cherished icon of the environmental movement. Could this be the same creature that William Hornaday, a leading conservationist in the early 1900s, described with these words: "Of all the wild creatures...none are more despicable than wolves. There is no depth of meanness, treachery or cruelty to which they do not cheerfully descend"? Clearly the differences in these perspectives say more about the people who hold them than they do about wolves. The personal value systems that underlie such perspectives play an important role in how policy and management decisions are made and whether people are willing to support these decisions. Being a conservation biologist means taking action to maintain the world's biological diversity, a prospect often fraught with controversy because of people's conflicting values. Conservation biologists need to consider social, political, and economic perspectives and to understand how people think and feel about the other life forms with which we share this planet.

People's values – what they hold as important – influence their choices about how to use the resources in the environment around them. People who value the wolf as an emblem of wilderness will have a very different perspective on wolf conservation plans than those who view the wolf as a threat to their property. Of course, there are many other points of view between these two extremes. When developing strategies for biodiversity conservation, core values as well as political, sociological, and economic factors are all important to consider. Taking these into account and understanding the stakeholders – people who will be directly or indirectly affected by a management option – are critical in developing a successful conservation strategy.

Researchers often use opinion polls or surveys to gather information about people's knowledge, attitudes, and behaviors. A survey is a systematic, impartial way of collecting information from a sample of people that represents a population. Results from surveys may be used in a variety of ways, such as to inform policy or to develop an appropriate educational campaign. Several national surveys have assessed the public's understanding of biodiversity and how people set priorities for loss of biodiversity in relation to other problems they face. Results from these surveys have been used to develop communication strategies appropriate for different audiences.

In the wolf example, survey-takers should be able to determine the proportions of residents who are for or against a wolf reintroduction project. They might also be able to tease out the basis for opposition to the project, as well as differentiate attitudes based on age or gender. This information would be very useful in developing a public outreach campaign for a wolf reintroduction project. Surveys can also be used to gauge the economic value that people place on biodiversity. For instance, a survey could be conducted to assess public support for a tax on camping goods that would be used to finance national parks or other reserves.

Administering a survey does not have to be prohibitively difficult, but it does take careful planning. This exercise takes you through the steps in conducting a poll:

1 Identifying a local or regional issue
2 Designing an opinion poll
3 Administering the poll
4 Analyzing the results.

At the end of the exercise you may not be an expert pollster, but by better understanding what goes into designing, implementing, and analyzing surveys, you can do a better job of assessing what others are saying when they tell you the results of their polls.

Objectives

- To learn about various public opinions concerning a biodiversity-related issue, to practice opinion-polling techniques
- To learn how to better assess the reliability of results reported from other surveys.

Procedures

Survey Objectives and Sample Population

The first step in writing and implementing a survey is to determine what information you want and what you will do with the results. As a class, choose a topic for the survey and decide what you would like to accomplish with the results. You may wish to use the biodiversity literacy survey questions in this exercise (see sample survey questions below). Otherwise, consider an issue that affects your university environment, such as the university's policies on recycling, energy use, or land development. What could you do with information about people's attitudes on these issues?

Once you have identified what you want to know, you need to identify whom to survey. For the wolf reintroduction example above, you would perhaps choose to interview residents within a 50-km radius of the proposed reintroduction area and select a random sample from this group. Concerning a tax on camping goods, you might want to poll subscribers to outdoor magazines or catalogs. For this class exercise, you should select a population and randomly sample 5 people (per poll-taker) from that population. For instance, to do a face-to-face survey of on-campus diners or people using the library (easy places to obtain a sample quickly), you can use a random numbers table to generate 5 numbers between 1 and 10 and take the sample from people passing by. For instance, if the numbers generated were 3, 5, 9, etc., you would interview the third person passing; when you complete this interview, count 5 people and interview the fifth, and so forth. If friends of the interviewee stop to listen to the survey, let them pass before you begin to count again. Some people may decline to participate. Decide the type of record of refusals you want to keep (e.g. noting gender, approximate age, reason for refusal, if any given). If the respondent who refuses was to be the fifth passerby, continue from that point by counting 5 people again and asking the fifth if he or she will answer your questions.

It is important to recognize that a sample of on-campus diners represents a subset of the student population at your college. In order to conduct a survey of the whole student body, you would need to select your sample from a phone book or other list that includes all students (and use a much larger sample size).

Survey Design

Surveys can be administered in written form or through an interview, either over the phone or face-to-face. In this exercise, the latter – in-person interviews – are recommended because you will be able to conduct the interviews in a relatively short period of time and with few resources (copies of the questionnaire, a clipboard, and a pen).

To prepare survey questions, it might be best to break the class into small working groups and have each group write enough questions so that all together the survey contains up to 10 questions. Coordinate which topics each group will cover with their questions.

The tone and scope of the questions in a survey can greatly influence the outcome. Some advocacy organizations take advantage of this and plan surveys to obtain pre-determined results. However, a good questionnaire should present questions in an unbiased and balanced way. A question is biased when one answer is obviously a better choice than another. In the following True/False example, the word choices reveal a bias which might influence respondents:

Killing helpless animals is acceptable if they are on your property.

"Killing" and "helpless" are both words that are emotionally charged. Sometimes bias introduced by word choice is more subtle, as in the statements below:

The government should **allow** private landowners to shoot wolves that enter their property.
The government should **forbid** private landowners from shooting wolves that enter their property.

When asked whether they agree or disagree, people may disagree with both statements if they wish to support wolf conservation, but also find it disagreeable that the government should "forbid" anything. When questioning people on their environmental behavior, they may have a tendency to answer based on what they feel is socially acceptable. It is important to try to craft questions to avoid this bias.

Survey questions about an issue should be balanced, meaning that both sides of the issue are presented with equal weight. For example, the following question (modified from Caro et al. 1994) briefly states positive and negative consequences of extensive tourism in national parks, allowing the respondent to weigh these in formulating an opinion:

> Extensive tourism in national parks can lead to new roads and facilities being built, erosion of hiking trails, and disturbance of animals. Should it be common park policy to limit tourist numbers in sensitive areas, even though this will decrease park revenue and hence economic incentives for maintaining parks?

All questions should pertain to the main focus of the poll. It is tempting to ask a question simply because it is interesting, but irrelevant questions waste time and energy. Always keep your overall purpose in mind: What are you trying to find out? What will you do with this information? Maintaining the connections between these aims and the survey questions is helpful when it comes time to analyze your results.

Effective questions are simple, clear, unambiguous, and generally short. While long questions can be difficult to answer, they may be necessary to address complex issues. Regardless of their length, each of your questions should focus on one idea and not ask two questions at a time. If a respondent were to disagree with the following statement, it would not be clear whether they disagreed with limiting the number of tourists in the national parks, with increased entrance fees, or both:

> Do you favor limiting the number of tourists to national parks by increasing park entrance fees?

Survey questions can be **close-ended** (with prescribed answers such as True or False, Agree or Disagree) or **open-ended** (where respondents answer in their own words). If well-written, close-ended questions are easy to analyze. Answer categories should be clear and easily distinguishable, such as: good, fair, poor. Categories like "fine" are not clear. Giving a "don't know" option avoids forcing people to state an opinion where they do not have one or are not familiar with the topic. The number of options for an answer should be between 2 and 6, but may go up to 8 or 9. However, if a list is too long, respondents are likely to choose answers near the top or bottom of the list.

Open-ended questions lend themselves to more detailed responses, allowing respondents to answer however they wish. Such responses are generally more difficult to analyze than close-ended questions. You may want to repeat an important question, once in an open form and once in a closed form, in different sections of the survey to check the respondent's consistency. Bear in mind that in any series of questions, a question may influence responses on subsequent questions; order can play a major role in determining your results. In general, questions should progress from general to specific, and if using open-ended questions, these should precede close-ended questions on the same topic.

Below are sample survey questions, question order, and demographic questions that may be used as guidelines as you design your survey.

Sample Survey Questions

The following are examples of the types of questions you might consider for a survey. Some of them ask about respondents' knowledge of a given subject, while others address attitudes.

Sample close-ended questions: True/False

1 Currently fewer than 100 species provide humans with the majority of their food supply.

True _____   False _____

2 Plants, birds, and large mammals make up about half of the world's species.

True _____   False _____

Sample close-ended questions: Yes/No

3 Have you ever heard about "the loss of biodiversity?"

Yes _____   No _____   Not sure _____

Sample close-ended questions: Agree/Disagree

Please tell me if you agree or disagree with each statement. (Answer choices may also be ranked as in question 7.)

4 It's often not worth the cost in jobs to try to save endangered species like spotted owls.

Agree _____   Disagree _____

5 The world would not suffer if some species, like poison ivy and mosquitoes, were eliminated.

Agree _____   Disagree _____

Ranked responses

6 From what you have heard or read, would you say the number of animal and plant species in the world is increasing, decreasing, or staying the same?

7 There need to be stricter laws and regulations to protect the environment.

_____ Completely agree
_____ Mostly agree
_____ Mostly disagree
_____ Completely disagree
_____ Don't know

8 Are you optimistic or pessimistic that the next generation will live in a better world than we do now in terms of the environment?

_____ Very optimistic
_____ Somewhat optimistic

_____ Somewhat pessimistic
_____ Very pessimistic
_____ Don't know/in between

9 What percentage would you be willing to pay in additional sales taxes on camping goods if this additional tax were used to buy land for parks?

_____ More than 5%
_____ Up to 5%
_____ Not willing to pay any extra tax

Sample open-ended questions

10 In your opinion, what is the most important environmental issue in your area?

11 In your opinion, what is the most critical threat to biodiversity in your area?

Sample Question Order

The following series of questions demonstrates a structure for a survey that includes questions pertaining to the knowledge and opinions of respondents regarding biodiversity. The questions are arranged so that they present information as respondents need it, and not before.

1 From what you have heard or read, would you say the number of animal and plant species in the world is increasing, decreasing, or staying the same?

2 Answer True or False: Plants, birds, and large mammals make up about half of the world's species.

True _____   False _____

3 Answer True or False: Fewer than 100 species currently provide humans with most of their food supply.

True _____   False_____

4 Have you ever heard about "the loss of biodiversity?"

Yes _____   No _____   Not sure _____

[Interviewer note; If "yes," continue with question 5. If "no" or "not sure," skip to question 6.]

5 In 15 words or less, explain what you think is meant by "the loss of biodiversity."

6 The term "biodiversity," short for biological diversity, refers to the variety of life in all its forms and the interactions among these living forms and their environment. In recent years, a debate has emerged between two groups of people: one group says we should be concerned because human activities are causing the loss of biodiversity; another group says that human activities such as economic development are more important than biodiversity conservation. Please state whether you completely agree, mostly agree, mostly disagree, completely disagree, or don't know.

There need to be stricter laws and regulations to protect biodiversity.
_____ Completely agree
_____ Mostly agree
_____ Mostly disagree
_____ Completely disagree
_____ Don't know

7 Are you optimistic or pessimistic that the next generation will live in a better world than we do now in terms of the environment?

_____ Very optimistic
_____ Somewhat optimistic
_____ Somewhat pessimistic
_____ Very pessimistic
_____ Don't know/in between

Sample Demographic Questions

One school of thought on survey design is that a survey should begin with questions that are relatively easy and non-threatening. Personal questions, while straightforward to answer, can put off a respondent. Therefore, it may be desirable to leave such demographic questions until the end when the respondent has had an opportunity to answer the substantive opinion questions.

> *Interviewer:* These last few questions are needed to compare your opinions with others and will help us analyze the results of this survey. Your answers will be kept completely confidential.

If anyone refuses to answer these questions, repeat the above statement. If he or she still declines, respect this wish and leave the responses blank, making a note that the respondent did not wish to answer.

> *Interviewer:* What is your age? _____

Some people are hesitant to state their exact age. You might provide choices of age ranges: less than 18 years, 18–25, 26–35, 36–45, 46–55 up to 86 or more. Make the scale appropriate for your survey sample (e.g. on a college campus, you may want to give ages in 5-year ranges between 18 and 35: 20 and under, 21–25, 26–30, 31–35), and so on.

> *Interviewer:* What is your sex? _____

This may well be obvious in a face-to-face interview, and could be simply recorded rather than asked.

> *Interviewer:* What is your profession?
> [Or (if a student), What is your field of study?]

Surveys often ask about level of education, income, and ethnic origin as well. Decide which of the above information would be useful to you in analyzing the results of your survey. Such demographic data can be helpful in identifying interest groups and determining critical audiences for information campaigns.

Survey Implementation

Once you have assembled a set of questions, practice asking it with your fellow students until you can get through the survey smoothly. Testing your survey will help you to identify questions that need to be revised because they are confusing, demand too much prior knowledge, or ask more than one thing at a time. As you practice the survey, record the time it takes to introduce it, read the questions, and record the answers. If it requires more than 10 minutes to administer the survey, consider eliminating or simplifying some of the questions.

When you administer the survey, identify yourself and the purpose of the survey, but avoid describing your goals in such a way that you influence how people respond. Assure respondents that their answers will be kept confidential. The following is a sample survey introductory statement to be given by the interviewer:

> "Hello, my name is _____. I am a student in the Department of _____. My Conservation Biology [or other] class is conducting a survey of public opinion on the environment. The survey takes only a few minutes and your answers will be kept strictly confidential."

In order to be consistent, always read questions exactly as they are written. Ask the respondents not to answer until you have finished reading the entire question. Try not to read too quickly; reading about two words per second gives respondents time to think about their response. Remember to ask all the questions consecutively and do not skip any.

Survey Analysis

When you have finished polling 5 people, collate the data on one sheet by question. For instance, for the question below, if 1 of the 5 respondents answered, "increasing," 3 answered, "decreasing," and 1 said, "staying the same," you would record data as follows:

> From what you have heard or read, would you say the number of animal and plant species in the world is increasing, decreasing, or staying the same?

Increasing	Decreasing	Staying the same
1	3	1

During the next class, tally your data with those of the other students and collate that data on one sheet (assuming 15 students conduct 5 surveys each, for a total of 75 respondents):

Increasing	Decreasing	Staying the same
26	30	19

Calculate percentages for each response to the questions. In the example above, 35% of the 75 respondents said "increasing," 40% responded "decreasing," 25%

responded "staying the same." Note: in a more comprehensive study, you would analyze the results by age, gender, field of study, or other descriptors to identify trends. For simple close-ended questions, you would use a chi-square analysis to compare results by gender or field of study. For close-ended questions that involve ranked responses or for open-ended questions, you would need to use non-parametric statistics to analyze your data.

After tabulating the results of your survey and calculating percentages of respondents giving a particular response, consider what these numbers mean. While polls can provide valuable information about what people know or feel, the relationships they reveal between opinions and particular traits, or between attitudes and actual behavior are generally weak. Be wary of overstating causal relationships. For example, if 60% of females correctly answered the survey True/False questions, while only 15% of males answered correctly, can you generalize that females are more knowledgeable about environmental issues than males (suggesting a correlation between gender and environmental knowledge)? What if 75% of the females you interviewed happened to be members of the campus environmental club, while none of the males were members?

Some questions to consider: What can you do with the information you have gathered with this survey? Who do you think should know the results? What actions should you, policy-makers, or other citizens take based on the opinions held by people you surveyed? Is there other information about people's values that you would like to find out?

Expected Products

- A printed, polling instrument, including a variety of question types, focused on a topic of interest as agreed upon by the class
- An analysis of the poll data as collected by class members who deployed the poll to an adequate sample of respondents
- A summary of the data analysis in terms of insights about the focal topic as gained from the polling exercise
- Responses in a form indicated by your instructor to the Discussion questions below.

Discussion

1 What does your experience through this exercise indicate about survey results in general and the conclusions that are reported in the media based on these results?

2 What can you do if you read about poll results that are contrary to what you consider the best course of action for biodiversity conservation?

3 Do you think what people say in a poll differs from how they act? If so, how would you account for this in interpreting polling results?

Making It Happen

Poll results are often reported in newspaper or journal articles, consumer reports, or newsletters from conservation organizations. Look for a poll on a topic that interests you. Try to obtain a copy of the questions to find out how the poll-takers arrived at their results and see if you agree with their conclusions. If you would like to get more involved in poll-taking, you could volunteer to work for a local policymaker or an environmental organization to canvas the public on environmental or biodiversity issues. If your university has an environmental club or a chapter of the Society for Conservation Biology, contact them to see if they need a poll done on a particular issue. You could ask polling experts to help you finalize your questions and to get more information on analyzing the results. Another interesting way to find out what issues concern people in your community is to attend a town council or natural resources commission meeting. Often such meetings provide a forum for people to express their opinions on a variety of issues. One thing to remember is that most institutions have specific policies regarding research on human subjects. If you undertake research using surveys or questionnaires, you will want to check on the Institutional Review Board requirements at your institution.

Further Resources

Several books have been written on people's attitudes towards biodiversity; a good one is Perlman and Adelson (1997). For examples of studies measuring people's attitudes towards a specific conservation issue using opinion polls, see Reading and Kellert (1993) and Shrestha and Alavalapati (2006). A fascinating look at contemporary attitudes toward biodiversity is Belden Russonello & Stewart's poll "Americans and biodiversity: New perspectives in 2002" available at www. biodiversityproject.org/02toplines.pdf.

CHAPTER 29

Priority Setting: Where Around the Globe Should We Invest Our Conservation Efforts?

James P. Gibbs

In the global effort to curb losses of biological diversity, many international conservation organizations try to set priorities for which regions should receive the relatively scarce resources available for conservation efforts. Organizations may set priorities for resource investment at global and regional, as well as local scales. Since it is impossible to measure and monitor all biodiversity even at a local scale, they choose surrogates as criteria for setting priorities. Common criteria include species richness (total number of species in an area), number of endemic species, and level of threat to species. They generally select species for which there is adequate data (mostly larger vertebrates) and assume that these surrogates represent other aspects of an area's biodiversity.

Areas chosen for global conservation effort generally have unusually large numbers of species, particularly endemic ones. Conservation International identifies global "hotspots" where species are numerous, often endemic, and under threat. Examples include the Caribbean Islands, the Caucasus Mountains, Madagascar and the Indian Ocean Islands, New Caledonia, Polynesia-Micronesia and many other regions (Figure 29.1).

How does one go about the process of identifying countries or regions in terms of relative immediacy of need for resources and assistance? It can be a complex undertaking, but many of the techniques used can be applied at the global, regional, or local level. One attempt was made by Dinerstein and Wikramanayake (1993) who contrasted biodiversity security with threat to prioritize 23 Indo-Pacific countries (Figure 29.2). More specifically they used estimates of the extent of protected areas (in some sense "biodiversity security") and deforestation (in some sense "biodiversity threat") to classify each country into four categories: well protected but with little forest outside protected areas (category I), well protected and with extensive forest remaining outside protected areas (category II), poorly protected but with extensive forest outside protected areas (category III), and poorly protected and with little forest outside protected areas (category IV, Figure 29.2). Category IV countries, i.e. those with little protected area and high rates of habitat loss, might qualify as those requiring the most urgent action. Conversely, category I

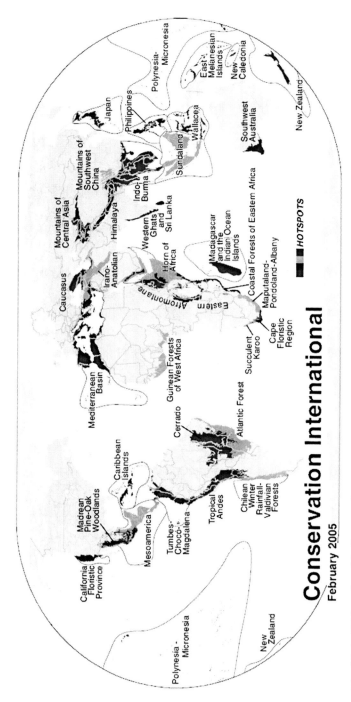

Conservation International

February 2005

HOTSPOTS

Fig. 29.1 Thirty-four biodiversity hotspots identified by the conservation organization Conservation International and used in that organization's conservation planning efforts.
Source: Conservation International: www.biodiversityhotspots.org.

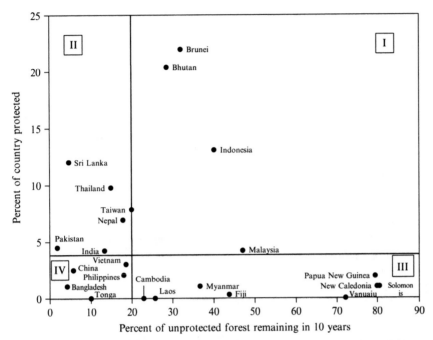

Fig. 29.2 Approach used by Dinerstein and Wikramanayake (1993) to prioritize conservation efforts among 23 Indo-Pacific countries. Countries were placed into four categories as follows: I. Countries with a relatively large percentage (>4%) of forests under formal protection and that will have a high proportion (>20%) of unprotected forested areas left in 10 years; II. Countries with a relatively large percentage of forest (>4%) under formal protection, but that will have little (<20%) unprotected forests left in 10 years; III. Countries with a relatively low percentage (<4%) of forests presently protected and that under current deforestation rates these countries will still retain a large proportion (>20%) of their unprotected forests in 10 years; IV. Countries with a relatively low proportion (<4%) of forests presently protected and little forest remaining.

countries might seem to require the least attention and category II and category III countries were best candidates for moving to category I status as quickly as possible. This analysis could also be expanded to include other considerations, such as the amount of corruption prevalent in a particular country or region, that country or region's political stability and its level of economic development. Similarly, the amount of diversity found in each country might also have been evaluated, that is, by considering the number of endemic species present as well as those potentially or actually at risk.

Different organizations and individuals will approach this complicated prioritization problem differently depending on their objectives and constraints. It is much more than an academic exercise; these prioritization efforts affect where funding and assistance will flow with tangible effects on people and wild species in different parts of the world. What would you do if you had to make these decisions?

Objective

- To identify the top priority "hotspots" for allocating resources for conservation.

Procedures

Hotspot Data

In order to explore one of the global priority setting strategies, current data on 34 global biodiversity "hotspots" as identified and compiled by Conservation International are provided (Table 29.1). The data provide a rich set of information about each hotspot and include: its original extent, how much of the original vegetation of the hotspot remains, the estimated number of endemic plant and endemic threatened bird, mammal, and amphibian species found in the hotspot, the number of recorded species extinctions in the hotspot, the human population density there and estimates of the aggregate area protected as well as the aggregate area in types of protected areas categories regarded by the IUCN as being afforded higher levels of protection (Categories I–IV). (An electronic form of these data is provided for you to download at this book's website.)

The Problem

Your task is relatively simple – identify the top priority "hotspots" for allocating resources for conservation. How will you carry out this task? There are many ways to approach this problem *and there is no single or correct answer.*

You need to devise some form of ranking system; the rationale for the approach you choose should be outlined and articulated clearly. Start, then, with conceptualizing what factors you think are most indicative of impending threats to biological diversity in a given "hotspot." Also consider the intrinsic value of a particular "hotspot" in terms of the biodiversity it harbors. What are the critical factors to consider?

Establishing Ranking Criteria

Once you have identified relevant criteria to consider, how will you contrast "hotspots" in relation to these criteria? You should do this with an objective, data-based approach with your rationale clearly articulated. Note that in comparing among "hotspots" you may need to synthesize some new variables. Once you have identified your primary variables, how will you analyze the data to derive robust indicators that are also relatively few in number so that they can be communicated easily to your audience? Dinerstein and Wikramanayake's (1993) efforts depicted in Figure 29.2 are based on a two-axis, four-category system that is simultaneously a useful analytical and communication device. For starters, you might similarly identify just two primary axes and follow their approach. But if you are feeling ambitious you might consider three or more axes along which to array each "hotspot." Remember you will need to defend your approach and show that you used all available data to best effect.

Manipulating the Data

One approach to making sense of the large amount of data presented in Table 29.1 is to use criteria matrices to set priorities. Several methods for setting priorities have been developed that use various criteria. Most of these systems combine criteria of rarity/richness and threat. You can design your own system, identifying criteria that

Table 29.1 The status and attributes of 34 global biodiversity hotspots as defined by the conservation organization. Conservation International (source: www.biodiversityhotspots.org)

	Hotspot original extent (km²)	Hotspot vegetation remaining (km²)	Endemic plant species	Endemic threatened birds	Endemic threatened mammals	Endemic threatened amphibians	Extinct species (recorded since 1500)	Human population density (people/km²)	Area protected (km²)	Area protected (km²) in Categories I–IV
Caribbean islands	229,549	22,955	6,550	48	18	143	38	155	29,605	16,306
Caucasus	532,658	143,818	1,600	0	2	2	0	68	42,721	35,538
Madagascar and the Indian Ocean islands	600,461	60,046	11,600	57	51	61	45	32	18,482	14,664
New Caledonia	18,972	5,122	2,432	7	3	0	1	11	4,192	497
Polynesia–Micronesia	47,239	10,015	3,074	90	8	1	43	59	2,436	2,088
Cape Floristic Province	78,555	15,711	6,210	0	1	7	1	51	10,859	10,154
Eastern Melanesian islands	99,384	29,815	3,000	33	20	5	6	13	5,677	0
Eastern Afromontane	99,384	29,815	3,000	33	20	5	6	13	5,677	0
Succulent Karoo	102,691	29,780	2,439	0	1	1	1	4	2,567	1,890
Western Ghats–Sri Lanka	189,611	43,611	3,049	10	14	87	20	261	26,130	21,259
Mountains of Southwest China	262,446	20,996	3,500	2	3	3	0	32	14,034	4,273
New Zealand	270,197	59,443	1,865	63	3	4	23	14	74,260	59,794
Maputaland–Pondoland–Albany	274,136	67,163	1,900	0	2	6	0	70	23,051	20,322
Tumbes-Choco-Magdalena	274,597	65,903	2,750	21	7	8	4	51	34,338	18,814
Coastal forests of Eastern Africa	291,250	29,125	1,750	2	6	4	0	52	50,889	11,343
California Floristic Province	293,804	73,451	2,124	4	5	8	2	121	108,715	30,002
Phillipines	297,179	20,803	6,091	56	47	48	2	273	32,404	18,060
Wallacea	338,494	50,774	1,500	49	44	7	3	81	24,387	19,702

(*Continued*)

Table 29.1 (*Continued*)

Southwest Australia	356,717	107,015	2,948	3	6	3	2	5	38,379	38,258
Japan	373,490	74,698	1,950	10	21	19	7	336	62,025	21,918
Chilean winter rainfall–Valdivian forests	397,142	119,143	1,957	6	5	15	0	37	50,745	44,388
Madrean pine-oak woodlands	461,265	92,253	3,975	7	2	36	1	32	27,361	8,900
Guinean forests of West Africa	620,314	93,047	1,800	31	35	49	0	137	108,104	18,880
Himalaya	741,706	185,427	3,160	8	4	4	0	123	112,578	77,739
Mountains of Central Asia	863,362	172,672	1,500	0	3	1	0	42	59,563	58,605
Irano-Anatolian	899,773	134,966	2,500	0	3	2	0	58	56,193	25,783
Mesoamerica	1,130,019	226,004	2,941	31	29	232	7	72	142,103	63,902
Sundaland	1,501,063	100,571	15,000	43	60	59	4	153	179,723	77,408
Tropical Andes	1,542,644	385,661	15,000	110	14	363	2	37	246,871	121,650
Horn of Africa	1,659,363	82,968	2,750	9	8	1	1	23	145,322	51,229
Cerrado	2,031,990	438,910	4,400	10	4	2	0	13	111,051	28,736
Mediterranean Basin	2,085,292	98,009	11,700	9	11	14	5	111	90,242	28,751
Atlantic forest	1,233,875	99,944	8	55	21	14	1	87	50,370	22,782
Indo-Burma	2,373,057	118,653	7	18	25	35	1	134	235,758	132,283

are most important in your view. A matrix approach is most useful when dealing with a large number of criteria to be incorporated when you wish to weight each criterion individually.

Here is an example of a criteria matrix to set priorities using data for a subset of just three of the "hotspots" listed in Table 29.1. Let's assume that you have decided that your primary axes of interest are the degree of "threat" and intrinsic "importance" to global biodiversity. For "threat" you decide you are most concerned with the percentage of the original vegetation remaining that is strictly protected (something you calculate from Hotspot Vegetation Remaining (km²) and Area Protected (km²) in Categories I–IV in Table 29.1) as well as human density (assuming that increasing numbers of humans pose a greater threat). You might also assume these two factors are of equal consideration so you apply an equal weight to them (here = 1, but it can be any equivalent values). The actual ratings of each variable times the importance weight you assigned ($r \times w$) summed together equals the overall score for "threat." A similar process is applied to "importance." In this case, you have assumed that endemic plant species richness is of equal importance to that represented by all threatened endemic vertebrates (birds, mammals and amphibians) so you apply a weight of 1 for plants and 0.333 for each vertebrate group (you could have also applied 3 to the plants and 1 to each vertebrate group and gotten the same results). Make sure in all cases that your weight is consistent with what you wish it to reflect, e.g., higher weights should indicate greater importance or threat. Summing across the rxw scores for each criterion you get an over "importance" score for each "hotspot" relative to the others "hotspots." Table 29.2 indicates how this is done for a small subset of sites.

Table 29.2 Sample calculations as an example of a criteria matrix to set priorities using data for a subset of just three "hotspots" listed in Table 29.1; for data that have not been converted to fractions of the maximum value observed in any hotspot.

Criterion	Criterion weighting	Caribbean islands		Caucasus		Madagascar and the Indian Ocean islands	
		Rating	w × r	Rating	w × r	Rating	w × r
Threat index							
% Remaining vegetation strictly protected[1]	1.0	71.0	71.0	24.7	24.7	24.4	24.4
Human population density (people/km²)	1.0	155.0	155.0	68.0	68.0	32.0	32.0
Total threat score			**226.0**		**92.7**		**56.4**
Importance index							
Endemic plant species	1.0	6 550	6 550.0	1600	1 600.0	11 600	11 600.0
Endemic threatened birds	0.3	48	16.0	0	0.0	57	19.0
Endemic threatened mammals	0.3	18	6.0	2	0.7	51	17.0
Endemic threatened amphibians	0.3	143	47.6	2	0.7	61	20.3
Total importance score			**6 619.6**		**1 601.3**		**11 656**

[1] calculated from Hotspot vegetation remaining (km²)/Area protected (km²) in Categories I–IV×100.

As a final recommendation, we suggest dividing the values from Table 29.1 by the maximum value observed in any given hotspot so that data are converted to fractions (0–1) of the maximum. For example, divide endemic threatened birds for all hotspots by 110 (the maximum value observed [in the Tropical Andes]). This way all parameters will vary by the same amount and your weights can then function properly. (If you didn't do this conversion then, for example, the number of plant species will overwhelm the other parameters simply because it represents the largest values and would influence the sum of the products the most). This approach is portrayed in Table 29.3.

Last, you can plot these scores for the three countries in a bi-variate plot as did Dinerstein and Wikramanayake (1993) and you get Figure 29.3.

The final step is to apply your ranking system to all 34 hotspots. You can eventually identify the category IV countries by determining the median value of all hotspots for each axis and determining which hotspots fall below the median values for each axis you evaluate. This is a convenient way to categorize "hotspots" and identify the priority ones with the category systems used by Dinerstein and Wikramanayake (1993). You may well decide on another approach. Any approach is fine as long as it is logical, it is well documented, and it stands up to review by your peers.

Expected Products

- A presentation (in written, verbal, or presentation form as your instructor prefers) of the rationale and associated methods you used for prioritizing the 34 hotspots
- A list and short description of the priority hotspots that you identified and explanation of why they merit this distinction
- Responses in a form indicated by your instructor to the Discussion questions below.

Discussion

1 What other kinds of data needs might improve the ranking process? Where would you secure them?
2 Were the same hotspots prioritized by all parties working on this problem? If, not, why not? Whose approach is best?
3 Is the "hotspots" approach the "silver bullet" strategy for conserving most species for least cost? Are all taxa equally represented? Can a few well-studied groups of higher plants and vertebrate animals serve as surrogates for the many other groups not included in these assessments?
4 What do you think should be the priority criteria: where species diversity is greatest, areas faced with imminent destruction, or large intact ecosystems? Or should we not prioritize at all given that nature everywhere benefits from conservation?
5 What is the next step in terms of getting your results integrated into the policy process? How actually do international conservation groups implement these kinds of analyses?

Table 29.3 Sample calculations as an example of a criteria matrix to set priorities using data adjusted as proportions for a subset of just three "hotspots" listed in Table 29.1.

Criterion	Criterion weighting	Caribbean islands		Caucasus		Madagascar and the Indian Ocean islands	
		Rating	w × r	Rating	w × r	Rating	w × r
Threat index							
% Remaining vegetation strictly protected[1]	1.0	1.0	1.0	0.3	0.3	0.3	0.3
Human population density (people/km^2)	1.0	1.0	1.0	0.4	0.4	0.2	0.2
Total threat score			**2.0**		**0.8**		**0.6**
Importance index							
Endemic plant species	1.0	0.6	0.6	0.1	0.1	1.0	1.0
Endemic threatened birds	0.3	0.8	0.3	0.0	0.0	1.0	0.3
Endemic threatened mammals	0.3	0.4	0.1	0.0	0.0	1.0	0.3
Endemic threatened amphibians	0.3	1.0	0.3	0.0	0.0	0.4	0.1
Total importance score			**1.3**		**0.2**		**1.8**

[1] calculated from Hotspot vegetation remaining (km^2)/Area protected (km^2) in Categories I–IV×100.

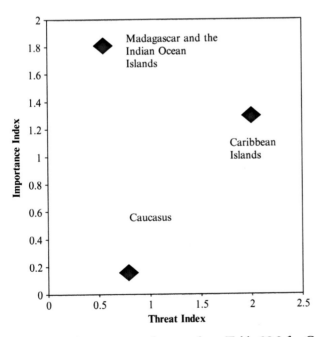

Fig. 29.3 Final "threats" and "importance" scores from Table 29.2 for Caribbean islands, Caucasus, and Madagascar and the Indian Ocean islands, portrayed as a bi-variate plot.

Making It Happen

Prioritization of regions at the global scale is done primarily to guide to policy-makers and conservation financiers. This said, the concepts of site prioritization outlined here still apply at the local level where an individual can be quite effective. In your region (e.g. county, province, etc.) where do you think conservation resources would best be expended? How would you tackle the problem on a local level? On a related note, these approaches to prioritization apply in many contexts in conservation biology, what species to emphasize or what ecological indicators to monitor, and the approaches outlined here will apply in those contexts as well.

Further Resources

Two good overviews of the "hotspots" concept are Myers et al. (2000) and Reid (1998) and one applied to the marine realm (Roberts et al. 2002). A recent elaboration of some of the challenges of identifying biodiversity hotspots for multiple taxa is Oertli et al. (2005). For an overview of Conservation International's attempt to inventory and classify biodiversity hotspots around the world, see: www.biodiversityhotspots.org.

CHAPTER 30

An International Debate: Commercial Fishing in Galápagos National Park

Malcolm L. Hunter, Jr., and Krishnan Sudharsan

Dusty cones of lava and scrub are not the most photogenic of landscapes, but the Galápagos Islands draw thousands of tourists every year. Some come because this is the archipelago where island-to-island variation in the forms of mockingbirds, tortoises, and finches catalyzed Darwin's formulation of the theory of evolution. More come because the extraordinary inhabitants of these islands – the giant tortoises, marine iguanas, penguins, and more – are as captivating for visitors today as they were for Darwin. Unfortunately, the Galápagos are far from being an untouched paradise. Early mariners ransacked the tortoise populations and left rats, goats, and other species that still plague many of the islands. More recently, the world's conservationists have turned their attention to the islands because of another threat: large numbers of people coming to the islands. Some come as tourists with the best of intentions but they inevitably bring issues such as solid waste disposal, sewage, and invasive stowaways. Others come planning to stay, attracted by economic opportunities in fisheries and the tourism business. Resolution of these problems has not been easy and has lead to considerable conflict among long-term residents of the islands, newcomers, the national government of Ecuador, and other countries that wish to see the Galápagos remain the global treasure that it is. All in all, we seem to be at a critical juncture in the history of these remarkable islands.

Objectives

- To learn about the problems facing the Galápagos islands
- To experience, through role-playing, how these issues are likely to be debated in an international forum.

Procedures

The basic format for this exercise will be a debate of the World Heritage Committee (WHC), which is a part of Unesco (the United Nations Educational, Scientific, and Cultural Organization) that fosters protection of the world's most important cultural and natural features (e.g. the Taj Mahal and Yellowstone National Park). Galápagos National Park is an official World Heritage Site and this gives the WHC some oversight on activities there. The specific issue to be debated is whether or not overfishing and associated problems are so severe that the WHC should put the Galápagos National Park on its official List of World Heritage in Danger. For this debate you will assume the role of a delegate to the committee representing one of several countries, or a representative of a nongovernmental organization (NGO).

To prepare for this exercise you need to do four things before class:

1 Read some background material about the issues. You can find a short summary at: www.galapagos.org/conservation.html, a book-length treatment in D'Orso (2002), and various specific articles such as Camhi (1995), Cruz et al. (2005), and Gottdenker et al. (2005). To understand how the World Heritage Committee works, refer to whc.unesco.org where you will some background material on World Heritage in Danger under the tab named "LIST."

2 Read some background information about the country that you represent. Your library will probably have some good material; if not, you can find the basics on-line through www.wikipedia.org. Also find something to wear that is emblematic of the country you are representing (e.g., a baseball cap for the US or a beret for France). This will greatly facilitate keeping track of who is from where.

3 Familiarize yourself with the list of participants and their agendas. Note which people are likely to vote with you, which will probably be against you, and which may be open to being influenced. (People will be wearing name tags so do not worry about not knowing everyone.)

4 Turn your agenda into about 60 seconds of succinct, persuasive, and eloquent oratory because the session will begin with each delegate giving a very brief summation of their country's position. Try to think of a unique argument that will make your presentation stand out in the crowd. Your instructor will also give you a hidden agenda that only you will know. For example, your secret instruction may remind you that while you have an official position to clearly articulate on this issue you also might need to tread carefully because you need political assistance on an important, seemingly unrelated matter, e.g. a trade issue, from another country that is taking a different position. This hidden agenda may shape your approach to this undertaking in many different ways. In some cases, it will affect the specific points you wish to raise in your formal presentation; in some cases, it will only affect your informal negotiations; in some cases, it has little to do with the issue but will still affect the way you operate.

On the day of the debate the session will be divided into three parts. Initially each delegate will give a brief speech. After they are finished, representatives of other organizations will be allowed to speak briefly as well.

After the opening speeches, you will adjourn and engage in an informal social hour during which your objective will be to convince other delegates to adopt your position. It would be wise to have thought out your strategy ahead of time, and remember, people are not always what they seem to be.

Finally, the group will reconvene and take a roll call vote to support or oppose the proposal. Only delegates to the WHC, not other interested parties, may vote.

A *final, important note* The agendas and hidden agendas used here are totally fictitious. In particular, many delegates have been given hidden agendas that are less than honorable because this will make the exercise more interesting and fun. The real members of the WHC, as well as citizens in general of the nations you will represent, act far more honorably than many of you will during this exercise.

Agendas

Algeria's agenda

As a Mediterranean nation Algeria has seen what can happen to marine resources when they are not properly managed and thus is inclined to send a message to Ecuador that it needs to do a better job of conserving the resources of the Galápagos.

Australia's agenda

Australia feels rather torn on this issue. Your management of the Great Barrier Reef (another World Heritage Site) is exemplary and you feel a need to be at the forefront of marine conservation. On the other hand, you are trying to loosen your links to the British Commonwealth and become more tightly connected with other Pacific Rim nations, especially the Asian countries that provide the main market for the Galápagos fisheries.

Bolivia's agenda

As a landlocked nation, Bolivia has only limited interest in protecting access to fisheries and maintaining maritime sovereignty. It is also actively developing some major, new national parks with substantial assistance from the World Wide Fund for Nature. Both of these factors tilt you toward voting for listing the Galápagos National Park as in danger. On the other hand, you are reluctant to vote against the interests of another Latin American country.

Canada's agenda

Since signing the Biodiversity Treaty from UNCED (the Earth Summit) in Rio de Janeiro, Canada sees itself as a leader among nations in international conservation efforts and feels compelled to protect the integrity of the WHC process by voting in favor of listing.

Colombia's agenda

Colombia wishes to play the part of a true neighbor by supporting the Government of Ecuador. It is also widely known that you are a close friend of the delegate from Ecuador, having been college roommates.

Democratic Republic of Congo's agenda

The DRC (formerly Zaire) believes that the standards for listing a World Heritage Site as in danger need to be revised. There are too many sites being listed or considered for listing because delegates from rich nations are unsympathetic to the problems faced by the governments of developing nations.

Ecuador's agenda

Ecuador is very embarrassed by the prospect of having one of it most famous features listed as being in danger. You are also offended that outsiders would take this opportunity to meddle in your affairs when you are doing your best to resolve an extremely difficult problem.

France's agenda

France supports the view that the people of Ecuador must be given other viable options before the Galápagos is placed on the List of World Heritage in Danger. You feel that the economy of Ecuador should be protected and since many Ecuadorians depend on fishing and on tourism for their livelihoods any unreal sanctions could have adverse effects. France does not want to impose any sanctions on Ecuador but instead, would like to see a plan to protect biodiversity as well the economy.

Germany's agenda

As President of the WHC you will not vote except to break a tie. Your agenda is simply to assure a fair and open debate on this issue.

Ghana's agenda

Ghana, being a small country similar in size to Ecuador, would like to maintain the longstanding friendship the two countries have had. With all the First World nations putting pressure on developing nations you reason that there is strength in numbers. By supporting Ecuador now, you hope to have an ally if problems of an international nature develop in your own country.

Japan's agenda

Japan has long been dependent on fisheries products gathered from throughout the world's vast oceans to feed the people of a crowded island, and thus is absolutely opposed to any effort that might curtail the Galápagos fisheries and have ramifications for fisheries management in many other nations.

Mexico's agenda

Mexico will support the WHC proposed listing because you strongly feel that it is critical that the world's heritage be protected. Your nation has been a highly responsible guardian of many WHC cultural sites, principally Aztec and Mayan ruins, and you feel that Ecuador should emulate your fine example.

North Korea's agenda

North Korea has a major role in the marketing of shark fins and other fishery products in Asia. Purchasing the commodities in Ecuador and then transporting them to key distribution centers in Taiwan and China has become a major source of hard currency for your faltering economy. Your people need food imports but will not be able to pay for rice without more hard currency.

Peru's agenda

You have spent a few summer vacations on the Galápagos islands and find its fauna and flora to be fascinating. To protect this island's treasures you support the move to place it on the World Heritage in Danger list. Of course, it is no secret that Peru and Ecuador have engaged in many border disputes and thus no one would expect you to support Ecuador.

Russia's agenda

Russia opposes the decision to place the Galápagos Islands on the List of World Heritage in Danger. Russia claims to understand the plight of developing countries trying to compete in the world market and feels that conservation should not be in the way of economic development. You are of the opinion that Ecuador has taken necessary steps to prevent the overexploitation of its resources and the real problem lies with international fishing vessels breaking the law. Besides, as a retired general you are still more interested in military issues than conservation. It is more than obvious that trips to the WHC meetings are just a nice excuse for you to travel around the world at government expense.

South Africa's agenda

South Africa is open to suggestions from Ecuador for protective measures, but unless concrete steps are taken to ensure the safety of the island, you will be forced to support placing the Galápagos Islands on the List of World Heritage in Danger.

Sweden's agenda

The Swedish Aid agency has a major project to work in the Galápagos fishing communities improving the welfare of the people and making their harvests more sustainable in the long run. You feel that a vote to place the park on the list of sites in danger may jeopardize all of your hard work to foster a local solution.

United Kingdom's agenda

As an island, seafaring nation you are extremely sensitive to the need to protect marine resources and the difficulties of doing so while providing for the needs of a large human population. However, while you are sympathetic to Ecuador's dilemma you must think first of the interests of all the world's people and protect our global heritage.

United States of America's agenda

The US actively supports the view that the Galápagos Islands should be placed on the List of World Heritage in Danger. The US would also like to provide any naval assistance required in keeping international fishing vessels out of Galápagos waters to protect marine ecosystems.

Agencia Turismo Ecuatoriana's agenda

Your organization is adamant that something be done to curb the abuses of the natural wonders of the Galápagos through overfishing, but you would not want to see the park listed as in danger because this is likely to discourage tourists from visiting.

Your organization has been working behind the scenes with the government of Ecuador, trying very hard to find a reasonable solution to this problem. Unfortunately, after years of things getting worse you reluctantly have to call upon conservationists throughout the world to send a clear message to the Government of Ecuador by listing the park as in danger.

World Wide Fund for Nature's agenda

You were working with fellow conservationists in Ecuador before there even was a Galápagos National Park and are extremely worried about your investment being ruined by recent developments. You feel that listing is the only prudent course and that doing so will send a message to countries throughout the world that they must protect their natural heritage.

Expected Products

- Demonstrated familiarity with Galápagos conservation issues based on the resources provided as well as those you unearth
- Demonstrated familiarity with the country that you represent
- Understanding of the political landscape of the debate, that is, familiarity with the participants and their agendas
- Presentation of succinct, persuasive, and eloquent oratory summarizing your country's position
- Response in a format indicated by your instructor to the Discussion questions posed below.

Discussion

1 What happened? Why did it happen? This is the time to reconstruct what went on: who persuaded whom and by what means; who had hidden agendas at odds with their open agenda; and so on.
2 Which parts of this exercise seemed relatively realistic? Which parts felt more like a role-playing exercise? What lessons did you learn that you could apply to analogous debates closer to home?

Making It Happen

Unless the next annual meeting of the World Heritage Committee is in a city near you it is unlikely that you would be able to attend and participate in one of their sessions. However, similar debates on a more local level happen all the time among state legislators, county commissioners, and various councils and boards. These meetings are almost always open to the public and very often there is an opportunity for the public to submit testimony. It is really eye-opening to attend one of these sessions and is especially good preparation for your professional career if you testify. The most important thing is that you project knowledge and sincerity. It is not all that hard to appear knowledgeable if you keep your presentation short and simple. Indeed, there is often a strict time limit that will compel you to stick to one or two key points.

Further Resources

Links to the websites mentioned above can be found at the website for this book. For the most relevant background reading on conservation in the Galápagos Islands, see D'Orso 2002.

CHAPTER 31

Conservation Law: Should the Polar Bear be Listed as a Threatened Species?

Thane Joyal

Administrative rule-making is arguably the most important mechanism affecting resource protection in many parts of the world with functional legal systems. This is where the "rubber meets the road," so to speak, in terms of how laws are interpreted and translated into mandates for action on the ground. Understanding the process by which the public – as individual citizens, conservation professionals, or lobbyists – can influence the making of rules that affect biodiversity is the first step in becoming an effective participant in the process.

To encourage public participation, the process itself is fairly straightforward. In the United States, for example, the formal procedures for rule-making in the federal Administrative Procedure Act are simple and easy to follow. However, the complex tensions between and within stakeholders in any administrative rule-making makes the process a vital and important forum for clarifying and resolving policy issues. Changes in listings under the Federal Endangered Species Act are a good example of regulations that involve diverse stakeholders in critical decision-making.

The purpose of this exercise is to introduce you to the basic procedures and tensions of the rulemaking process associated with a significant law protecting biodiversity. We focus on administrative rule-making under the US Administrative Procedure Act (5 USC sec. 500 *et seq.*) Note that the procedures outlined here are similar to those in related contexts around the world and the skills gained are generally quite transferable. As arcane as rule-making may seem, it is at the heart of the process by which laws are implemented. Any practicing conservation professional will encounter and likely even participate in the rule-making process at some point in their career.

Objectives

- To understand the procedures for rulemaking in the US Administrative Procedure Act

- To identify the parts of a proposed rule as published in the Federal Register
- To quickly review and understand the context of a proposed rule
- To identify key issues and appropriately comment on a proposed rule
- To engage in policy discussions concerning a proposed rule.

Procedures

Overview

You will read the January 9, 2007 rule entitled *Endangered and Threatened Wildlife and Plants: 12-Month Petition Finding and Proposed Rule to List the Polar Bear (Ursus maritimus) as Threatened Throughout its Range* at 72 F.R. 1064. This document is available at this book's website (see page ix). You will be given a particular issue to focus on within the rule. You will be asked to prepare a short worksheet on the rule before class and in the next class, you will be divided into small groups to conduct a simulated administrative "hearing" on the proposed rule. After each person has presented comments, Hearing Officers from each team will caucus and present a recommendation for each subject area within the rule, and an overall recommendation on the rule.

Group Assignments

You will be divided into groups of 5 to 7 students. One individual must be the hearing officer who will prepare the hearing officer worksheet. Anyone who is not a hearing officer is a "commenter." Commenters will select a particular topic (or set of topics) from those listed below:

Topic areas

- Distribution and movement
- Food habits
- Ice habitat relationships
- Current population status and trends
- Effect on prey
- Access to, and alteration of denning status
- Oil and gas exploration, development and production
- Harvest management by nation: Russia, Canada, Greenland, Norway, United States
- Polar bear – human interactions
- International agreements
- Mechanisms regulating sea ice recession
- Other natural or manmade factors affecting polar bear survival
- Available conservation methods.

You will review the proposed rule and complete the worksheet (Box 31.1) outside class.

Presenting Comments

To conduct this as a class exercise, the Hearing Officer should convene the hearing following the review period, explain the procedures, and open the hearing for

Box 31.1 Rule-making Practicum Worksheet for Commenters

What is the name of this proposed rule?

What is the deadline for comments?

What prompted the USFWS to initiate this rule-making?

What are the potential outcomes of this rule-making?

What are the procedures for commenting on this rule-making?

comment. Hearing Officers are responsible for maintaining order within their groups. Each commenter will typically be given no more than five minutes to speak on their issue and each is required to speak for at least three minutes (time-frames can be adjusted based on class size and available time). Hearing Officers will take notes on the issues presented and the substance of the comments. The worksheet overleaf (Box 31.2) will help you to prepare your comments.

Caucus and Presentation to the Director

Following the receipt of comments, Hearing Officers will caucus together for no more than ten minutes to discuss the comments on the rule. Your objective is to form a recommendation to present to the Director of the US Fish and Wildlife Service, a role typically played by the instructor. Hearing Officers should be prepared to explain to the Director the major areas of public concern about the rule; any alternative actions recommended by the public; and the recommendations of the Hearing Officers as to each category of issues raised by the public. One or more commenters will be selected to present their recommendation to the Director.

Framing the Issues

While Hearing Officers are reviewing comments, remaining individuals will be divided into 5 groups and asked to review the conclusion of the proposed rule as to each of the factors considered as bases for the listing decision:

- **Factor A** Present or threatened destruction, modification, or curtailment of the species' habitat or range (pp.1071–1081)
- **Factor B** Overutilization for commercial, recreational, scientific or educational purposes (pp.1081–1085)
- **Factor C** Disease and predation (pp.1085–1086)
- **Factor D** Inadequacy of existing regulatory mechanisms (pp.1086–1091)
- **Factor E** Other natural or manmade factors affecting the polar bear's continued existence.

You will review the rule and select a spokesperson to present the group's summary of (i) the USFWS considerations in evaluating the Factor; and (ii) the USFWS decision with respect to that Factor.

Final Presentations

To ensure that the entire class has a common understanding of this complex rule, before the Hearing Officers present their findings and recommendations, spokespersons for the Factor Groups should explain the USFWS decision as to each factor and the considerations involved in arriving at that decision. After these presentations, the instructor should answer any questions students have about the existing rule. When a common understanding has been reached, the Hearing Officers should present their findings and recommendations to the Director, who will keep a list of issue categories and recommendations on the board for discussion at the conclusion of the presentation. The Director should then evaluate the recommendations, and indicate his or her decision on the proposed rule.

Box 31.2 Comment Worksheet

What is your interest in this rule?

What are your qualifications to comment on this rule?

Are you commenting in favor of or in opposition to this rule?

Description of issue:

Comment on issue:

Recommended changes, if any:

Supporting data, if any:

In the event that this exercise is assigned not as a discussion and class-level simulation but as an in-depth project for students to complete individually, each student should undertake the exercise by completing the worksheet shown below (Box 31.3).

The proposal is based on the best information that the U.S. Fish and Wildlife Service staff has gathered concerning the subject matter, and has been approved by the Director of the Service. In the final analysis, the Director will need to decide what action to take: Should the polar bear be listed as threatened, as recommended in the proposal? What other actions might the Director take? The USFWS staff base the proposal on the 5 factors listed here:

- **Factor A** Present or threatened destruction, modification, or curtailment of the species' habitat or range (pp.1071–1081)
- **Factor B** Overutilization for commercial, recreational, scientific or educational purposes (pp.1081–1085)
- **Factor C** Disease and predation (pp.1085–1086)
- **Factor D** Inadequacy of existing regulatory mechanisms (pp.1086–1091)
- **Factor E** Other natural manmade factors affecting the polar bear's continued existence.

For each factor:

1 Identify the principal elements of the USFWS's decision-making
2 Evaluate the strengths and weaknesses of the arguments and data presented by the USFWS
3 Identify issues that are likely to be the subject of public comment.

Box 31.3　Worksheet for Non-participatory Exercise

What is the name of this proposed rule?

What is the law applicable to this proposal?

What will be the next step in the process?

Summarize your input by preparing a briefing paper for the Director of the US Fish and Wildlife Service, including the foregoing information. Based on your independent assessment of the likely public response to this proposal, should this proposal become a final rule?

Expected Products

- To have read and understand the proposed rule
- Preparation and presentation of a 1–2 paragraph summary of key issues *or* prepare a briefing paper (as appropriate to a class- or individual-based exercise)
- Responses in a form indicated by your instructor to the Discussion questions below.

Discussion

1 In this complex rule-making, the US Fish and Wildlife Service will likely receive comments from individuals from many backgrounds and with varying levels of expertise. Based on your particular interest or expertise, what is your evaluation of the proposal?
2 If you were a US Fish and Wildlife Service staff person, how would you advise the Agency to evaluate the comments it receives?
3 Do the applicable laws clarify or cloud the policy decisions facing the agencies?
4 How do international laws and treaties affect the Agency's decision-making?
5 If you conducted this as a class exercise, you should evaluate your role and the hearing officer's decision either orally or in writing. Was it realistic? Why and why not? How did the exercise affect your understanding of the rulemaking process?

Making It Happen

Although reading the legal notices and scanning the Federal Register for proposed rules of potential significance to readers is hardly the stuff of high drama and scintillation, there are many more engaging ways to keep abreast of current administrative actions raising conservation issues. Most United States administrative agencies have efficient, up-to-date websites that contain information and press releases about important initiatives. For example, for current issues involving endangered species, you should check the US Fish and Wildlife's endangered species webpage (see Further Resources). Reading current topical periodicals, attending scholarly meetings, and corresponding with colleagues is another way to learn about these issues. Public involvement is the critical step in the administrative process, either by attending a public information session or hearing, submitting comments on a proposed rule. You may wish to go a step further by informing others about the proposal, or participating in an established environmental group that engages in advocacy. Writing letters to the editor of your local newspaper about issues of concern, speaking to local groups, or simply circulating a proposal to individuals you believe to have expertise in the subject matter may all be effective in shifting the public policy debate on an issue. Remember, the most important step is to become informed. The second most important step is to meet all applicable deadlines for commenting if you decide to participate actively in the rule-making process. Note that many commentators believe that conservation professionals have an ethical obligation to participate in these important matters.

Further Resources

The proposed rule upon which this exercise is based is authorized by the United States Endangered Species Act, 16 USC 460, et seq., on the web at http://www.fws.gov/endangered/esaall.pdf (also see its implementing regulations at 50 CFR Part 402) and it should be consulted for the larger context of this exercise.

An useful summary of polar bear biology and population status is provided by Schleibe et al. (2006) in the "Range-wide Status Review of the Polar Bear (*Ursus maritimus*)," available at http://alaska.fws.gov/fisheries/mmm/polarbear/pdf/Polar_Bear_%20Status_Assessment.pdf

A good short introduction to the listing process under the Endangered Species Act and the issues raised by this proposed rule can be found in the Congressional Research Service Report for Congress, Polar Bear Listing As Endangered Species, at: www.opencrs.com/rpts/RS22582_20070125.pdf

Current issues involving endangered species are highlighted at the US Fish and Wildlife's endangered species webpage: www.fws.gov/endangered/.

CHAPTER 32

Conservation Policy: Shaping Your Government

Thane Joyal

"Give me liberty or give me death." The rhetoric of democracy can seem a bit overdone at times, but that jaded view usually comes from people who have enjoyed democracy for so long that they now take it for granted. Democracy is extraordinarily important for environmentalists because it provides a set of tools with which to shape the laws and institutions that underpin our society. Without democracy, powerful financial interests will prevail and society as a whole – and the biota as a whole – are likely to suffer. Democratic societies impose a responsibility on their citizens. We must be informed about issues and use our knowledge and values to shape our government. Biologists carry a double responsibility in this regard. First, we must act to do what is best for human society as a whole, not just for a few greedy individuals who would degrade the common environment for their own gain. Perhaps more importantly, we must speak for all the other creatures that share this planet but have no voice in human institutions.

In this exercise you will identify a current political issue affecting the conservation of biodiversity and to compose a letter concerning the issue to the editor of a local newspaper or to an elected official. Secondly you will identify a current proposed state or federal regulation that raises an issue related to the conservation of biodiversity and then, following the guidelines set forth below, to prepare comments on a chosen proposed regulation to the appropriate agency. It is important that you choose an issue and a proposed regulation that you feel strongly about so that you can comment upon them with both enthusiasm and some level of expertise.

Objective

- To learn how to participate in the making of public policy by contributing useful written or oral inputs.

Procedures

Finding Out What's Going On

The public policy debate takes place in many different forums. Issues related to conservation may be raised in a general, policy-making forum by a government agency, or they may arise as a result of a specific dispute concerning a single parcel of land. The best ways to keep track of what is going on are to read the local and national newspapers and to contact local and national conservation organizations. An important function of non-profit conservation and environmental organizations is to track the development of policy-making debates. Particularly useful are the newsletters and web pages of these organizations. For the US, a publication known as *The Conservation Directory*, published by the National Wildlife Federation and available at most public and university libraries, lists most major conservation organizations and government agencies, and is a good tool for identifying organizations with agendas that interest you. The internet is also an obvious source of timely information on conservation policy issues and developments.

In most democratic nations, the most important national conservation laws are adopted by an elected body of representatives and approved by a chief executive, such as a president or prime minister. These laws give an administrative agency, such as a Department of Natural Resources or similar agency the authority to develop the specifics of these programs. To do so, the agency is authorized to promulgate regulations to implement its statutory mandates. Regulations are generally proposed for public notice and comment prior to adoption, and all comments are considered before the final regulation is published. Often a public hearing is held to solicit comments on the proposed regulation.

The easiest way to track proposed regulations is to read the newsletters and web pages of environmental and conservation organizations. In the US, the proceedings and activities of federal administrative agencies can be tracked through the Federal Register, a publication which is available at large public libraries and at law libraries or on the internet at: www.gpoaccess.gov/fr/index.html. Any reference librarian should be able to show you to how to locate and use the Federal Register. The most useful feature of the Federal Register is that it publishes copies of proposed regulations so that you can read them for yourself and form your own opinions about their content. There are usually similar publications pertaining to state regulations. Again, your reference librarian or local conservation groups can orient you to regulatory issues at the state level.

Analyzing the Situation

Conservation biologists and wildlife managers typically are well-steeped in scientific expertise and culture. They often assume that their years of training in ecological theory, natural history, mathematical modeling, systematics, etc. entitle them to a privileged role in policy debates. If the scientist just provides all the facts to the debate, the reasoning goes, then satisfactory resolution of a conflict will follow. Furthermore, any scientist who enters the policy fray personally is often denigrated as being an "advocate" or as being "political."

To become effective participants in the policy process, and ultimately to meet with any success in resolving the biodiversity crisis, biologists need a more sophisticated understanding of the policy process. Policy analysis is not necessarily complicated but is unfamiliar to most scientists. Some insights on policy analysis that can assist a biologist to become a more effective participant are presented here:

1 Carefully examine one's own perspective. How do your beliefs, values, and expectations contrast with those of other participants in the same debate? Are there areas of common ground with your opponents? Are there areas of irreconcilable differences? Recognizing the underpinnings of your perspective will make you a much more effective participant.

2 Examine the problem carefully. What outcomes do you prefer? What alternative outcomes are available to you and the other participants? Would all alternatives be equally acceptable? Try to be specific about identifying and addressing the merits of all the alternatives available.

3 How do you expect a policy debate to proceed? Be aware that any policy develops through several, fairly predictable phases, including:

- a planning and information-gathering phase
- a debate phase where the alternatives are discussed
- an enactment phase where an alterative or combination of alternatives is selected
- an invocation phase where the rules agreed upon are put into practice
- an application phase where disputes with the enacted policy are resolved
- a review phase where assessment of the policy is undertaken
- finally a termination phase where the policy is revamped by moving back to the beginning of the process or is abandoned.

Preparing a "Letter to the Editor"

It is very common to read newspaper articles about situations that raise issues related to biodiversity, and it is also quite common to see that there are serious misunderstandings or misrepresentations of the science of conservation biology which are at the heart of the debate. What should a responsible conservation biologist do? What if there is an issue which is not even being raised by the newspaper? A letter to the editor enables you to inform a large audience about an issue. Keep in mind that the "letters to the editor" section of a newspaper is typically read more frequently than any other. Legislators and other important readers also often perceive letters to the editor as a display of public sentiment.

Letters to the editor can take many forms. They can explain how your issue relates to other items currently being covered in the news, correct facts after a misleading, inaccurate or biased letter or story, respond to other editorials or letters, reveal local ramifications of national issues or raise local public awareness of an issue.

Letters should be brief, clear, and to the point. Approximately one page, or about two to three short paragraphs (250–500 words total) should be all that is needed to adequately address an issue. Short letters are much more likely to be printed than long ones owing to newspaper space limitations. Be sure to know what you want to say and to say it plainly so people can understand. Be positive. Localize and personalize your story in order to make it relate more closely to the people whom you are trying to influence. Type or handwrite clearly and include your address. Make sure facts and figures are correct. If you are writing as part of a group (such as a conservation biology class) make sure to stagger submissions to maintain public awareness and interest over a longer period of time.

Writing to Elected Officials

You can have an impact on the formation of statutes by tracking interesting legislation and writing to your elected representatives to express your views on the proposals. Politicians do listen to their constituents. Letter writing is the most common form of communication with elected officials, especially members of Congress. Keep in mind that a large amount of mail flows into congressional offices and not all of it reaches the members. Nevertheless, well-written letters are at least seen by top staff personnel. Usually all letters submitted receive a response.

The best way to ensure that your letter is read is to keep it as brief and to-the-point as possible. Type your letters using your signature over your typed name at the end of your message. Alternatively, a carefully hand-written letter may have more of a personal touch and hence more impact (particularly in this age of mass mailings). Clearly identify your subject, including the name of the legislation and the bill number. State your reason for writing and explain how the issue would affect biodiversity conservation. Your own personal experience is your best supporting evidence. Remember you are not engaged in a debate, but are trying to persuade, so do not be argumentative. If you are a constituent, make sure you state this. Ask your representative to state a position on the issue in a reply. Timeliness is everything, and be particularly aware of the schedule for debate and voting on the issue you are writing about. Lastly, elected representatives mainly receive complaints from their constituents. Do not underestimate the value of praising an official for past activity that pleased you as a constituent. Similarly, if the official responds to your letter in a positive fashion, write to them again indicating your gratitude. This happens all too rarely.

Communicating with Government Agencies

The most important rules to remember in submitting comments to an administrative agency on a proposed regulation are simple. Meet the deadlines and follow all instructions in the public notice! If you fail to follow the instructions or to meet the deadline stated in the public notice, the agency will not be required to consider your comments.

Begin your comments by introducing yourself, stating who you represent, if you are commenting on behalf of an organization, or if you are commenting as an unaffiliated citizen. If you have any special expertise in a particular area, you should state it at the beginning of your comments. Comments should only address the issues raised by the proposal. This seems obvious, but it is quite common for the public to use the forum to raise pet concerns or issues, which greatly dilutes their effectiveness. Comments should be specific about the portion of the regulation being addressed. If there is a problem with, for example, a single word in paragraph 3 line 2, say so specifically. If the problem is theoretical or is an overall concern, say so. The most effective comments often lead with a discussion of theoretical or general concerns, and then follow with specific comments on the language of the regulation itself. If you see a problem and can recommend a solution in the form of revised language, do so.

Comments should be concise. One common practice among regulation writers is to summarize comments and group them with similar comments to enable decision-makers to evaluate the information presented. The more wordy a comment is, the

more likely it is to be misunderstood or misrepresented. Effective public participation depends in large part on good writing and speaking skills.

When submitting comments, you should request that a copy of the final regulation be sent to you, or, at a minimum, notice of its publication. You should also ask that you be notified of any public hearings scheduled to address the proposal. Most government agencies will routinely send both notices to all commenters, but it is a good practice to request the notices, both to ensure that you receive them and to communicate the seriousness of your interest in the regulation. A public hearing provides you with a further opportunity to make your opinions known.

If your comments are not heeded, take heart! Regulations are frequently revised as new programs implemented. If you have the patience to continue to track a newly adopted regulation, you may have the opportunity to comment on the revised version in the future. If you are directly affected by a newly adopted regulation, do not wait for the revision. Write to the administrative agency and detail the concern, following all the rules for formal comment writing. Effective lawmaking depends on good input from those affected by the laws.

Expected Products

- Identification of a current political issue affecting the conservation of biodiversity and to compose a letter concerning the issue to the editor of a local newspaper or to an elected official.
- Identification of a current proposed state or federal regulation that raises an issue related to the conservation of biodiversity and then prepare comments on a chosen proposed regulation to the appropriate agency.
- Responses in a form indicated by your instructor to the Discussion questions below.

Discussion

1 We frequently hear reference to the "science" of conservation biology. If conservation biologists are essentially scientists, would they not best serve the public by simply generating impartial information for the public debate and not getting directly involved in politics themselves? In other words, does not direct participation in the policy process taint the respectability of conservation biologists and thereby diminish their stature and usefulness to society?
2 Has your education been adequate to prepare you for effective participation in the policy process? What has been most useful? What has been lacking?
3 Why do so few members of the general public participate in the policy process? In other words, do you think it matters, one way or the other, if you submit a letter to a representative or comment on a proposed rule?

Making It Happen

This exercise is all about making it happen and is therefore an appropriate one to end on. The purpose of the exercise is to familiarize you with the process of getting your opinions entered into policy debates. The exercise will only have succeeded if you build upon this experience, generate letters and comments on a frequent basis, and become a regular participant in such debates. One of the greatest frustrations of many government regulators and elected officials is that they only hear from so-called special interest groups because the public at large generally does not take advantage of the many mechanisms available to it to participate in the debate. In other words, we encourage you to go ahead and mail your letters and comments.

Further Resources

We recommend Susan Jacobson's (1999) handbook for getting up to speed as a savvy, articulate and, eventually, effective conservation biologist.

Literature Cited

Amstrup, S. C., T. L. McDonald, and B. F. J. Manly (eds.) 2005. Handbook of capture-recapture analysis. Princeton University Press, Princeton, New Jersey.

Avise, J. C. and J. L. Hamrick (eds.) 1996. Conservation genetics: Case histories from nature. Chapman and Hall, New York.

Batary, P. and A. Baldi 2004. Evidence of an edge effect on avian nest success. Conservation Biology 18:389–400.

Beaumont, L. J., L. Hughes and M. Poulsen 2005. Predicting species distributions: Use of climatic parameters in BIOCLIM and its impact on predictions of species' current and future distributions. Ecological Modelling 186: 250–269.

Beissinger, S. R. and D. R. McCullough (eds.) 2002. Population viability analysis. University of Chicago Press, Chicago.

Benstead, J. P., P. H. de Rham, J.-L. Gattolliat, et al. 2003. Conserving Madagascar's freshwater biodiversity. BioScience 53:1101–1111.

Berland, L. 1955. Les arachnides de l'Afrique noire française. IFAN, Dakar.

Blanford, W. S. 1888. The fauna of British India, including Ceylon and Burma; Mammalia. Taylor & Francis, London.

Buckland, S. T. 2006. Point transect surveys for songbirds: Robust methodologies. Auk 123: 345–357.

Buckland, S. T., D. R. Anderson, K. P. Burnham, et al. 2001. Introduction to distance sampling: Estimating abundance of biological populations. Oxford University Press, Oxford.

Burnham, K. P. and D. R. Anderson 2002. Model selection and multi-model inference. Springer-Verlag, New York.

Camhi, M. 1995. Industrial fisheries threaten ecological integrity of the Galápagos Islands. Conservation Biology 9: 715–719.

Caro, T. M., N. Pelkey, and M. Grigione 1994. Effects of conservation biology education on attitudes toward nature. Conservation Biology 8: 846–852.

Carothers, A. D. 1973. Capture-recapture methods applied to a population with known parameters. Journal of Animal Ecology 42:125–146.

Caughley G. 1994. Directions in conservation biology. Journal of Animal Ecology 63: 215–244.

Caughley, G. and A. Gunn 1996. Conservation biology in theory and practice. Blackwell Science, Cambridge, Massachusetts.

Caughley, G. and A. R. E. Sinclair 1994. Wildlife ecology and management. Blackwell Science, Cambridge, Massachsetts.

Chao, A., R. L. Chazdon, R. K. Colwell, and T.-J. Shen 2005. A new statistical approach for assessing similarity of species composition with incidence and abundance data. Ecology Letters **8**: 148–159.

Chase, M. R., C. Moller, R. Kessell, and K. S. Bawa 1996. Distant gene flow in tropical trees. Nature **383**: 398–399.

Coddington, J. A. and H. W. Levi 1991. Systematics and evolution of spiders (Araneae). Annual Review of Ecology and Systematics **22**: 565–592.

Colwell, R. K. and J. A. Coddington 1994. Estimating terrestrial biodiversity through extrapolation. Philosophical Transactions of the Royal Society, London, Series B, **345**: 101–118.

Cooch, E. and G. White 2005. Program MARK: A gentle introduction. Available in .pdf format for free download at www.phidot.org/software/mark/docs/book/.

Costanza, R., R. d'Arge, R. de Groot, et al. 1997. The values of the world's ecosystem services and natural capital. Nature **387**: 253–260.

Cox, G. W. 2005. Conservation biology: Concepts and applications. McGraw-Hill, Dubuque, Iowa.

Cruz, F., C. J. Donlan, K. Campbell, et al. 2005. Conservation action in the Galápagos: Feral pig (*Sus scrofa*) eradication from Santiago Island. Biological Conservation **121**: 473–478.

Debinski, D. M., and R. D. Holt. 2000. A survey and overview of habitat fragmentation experiments. Conservation Biology **14**: 342–355.

Dinerstein, E. and E. D. Wikramanayake 1993. Beyond "hotspots": How to prioritize investments to conserve biodiversity in the Indo-Pacific region. Conservation Biology **7**: 53–65.

D'Orso, M. 2002. Plundering paradise: The hand of man on the Galápagos islands. Harper Collins, New York.

Elith, J., C. H. Graham, R. P. Anderson, et al. 2006. Novel methods improve prediction of species' distributions from occurrence data. Ecography **29**: 129–151.

Faith, D. P. 2002. Quantifying biodiversity: A phylogenetic perspective. Conservation Biology **16**: 248–252.

Falk, D. A., C. J. Millar, and M. Olwell (eds.) 1996. Restoring diversity: Strategies for reintroduction of endangered plants. Island Press, Washington, DC.

Fragoso, J. M. V. 1998. Home range and movement patterns of white-lipped peccary (*Tayassu pecari*) herds in the northern Brazilian Amazon. Biotropica **30**: 458–469.

Franco, M. and J. Silvertown. 2004. Comparative demography of plants based upon elasticities of vital rates. Ecology **85**: 531–538.

Frankham, R. 1995. Effective population size/adult population size ratios in wildlife: Review. Genetical Research **66**:95–107.

Frankham R., J. D. Ballou and D. A. Briscoe 2002. Introduction to conservation genetics. Cambridge University Press, Cambridge.

Gascon, C., T. E. Lovejoy, R. O. Bierregaard, et al. 1999. Matrix habitat and species richness in tropical forest remnants. Biological Conservation **91**: 223–229.

Gaston, K. L. 2000. Global patterns in biodiversity. Nature **405**: 220–227.

Geffen, E., M. J. Anderson, and R. K. Wayne 2004. Climate and habitat barriers to dispersal in the highly mobile grey wolf. Molecular Ecology **13**: 2481–2490.

Gotelli, N. J. 2001. A Primer of Ecology. 3rd edition. Sinauer Associates, Inc., Sunderland, Massachusetts.

Gotelli, N. J. and R. K. Colwell 2001. Quantifying biodiversity: Procedures and pitfalls in the measurement and comparison of species richness. Ecology Letters **4**: 379–391.

Gottdenker, N. L., T. Walsh, H. Vargas, et al. 2005. Assessing the risks of introduced chickens and their pathogens to native birds in the Galápagos Archipelago. Biological Conservation **126**: 429–439.

Govindasamy, D., P. B. Duffy, and J. Coquard 2003. High-resolution simulations of global climate, part 2: Effects of increased greenhouse gases. Climate Dynamics **21**: 391–404.

312

Graham, C. H. and R. J. Hijmans 2006. A comparison of methods for mapping species richness. Global Ecology and Biogeography **15**: 578–587.

Graham, C. H., S. Ferrier, F. Huettman, et al. 2004. New developments in museum-based informatics and application in biodiversity analysis. Trends in Ecology and Evolution **19**: 497–503.

Groom, M. J., G. K. Meffe, and C. R. Carroll 2006. Principles of Conservation Biology. Sinauer Associates, Sunderland, Massachusetts.

Guerrant, E. O., Jr., K. Havens, and M. Maunder (eds.) 2004. *Ex Situ* plant conservation supporting species survival in the wild. Island Press, Covelo, California.

Hall, P., M. R. Chase, and K. S. Bawa 1994. Low genetic variation but high population differentiation in a common tropical forest tree species. Conservation Biology **8**: 471–482.

Hanson, E. 2002. Animal attractions. Princeton University Press, Princeton, New Jersey.

Hartley, M. and M. L. Hunter, Jr. 1998. A meta-analysis of forest cover, edge effects, and predation rates of artificial nests. Conservation Biology **12**: 465–469.

Hijmans, R. J., S. E. Cameron, J. L. Parra, et al. 2004. The WorldClim interpolated global terrestrial climate surfaces. Version 1.3. Available online at http://biogeo.berkeley.edu.

Hijmans R. J. and D. M. Spooner 2001. Geographic distribution of wild potato species. American Journal of Botany **88**: 2101–2112. Online at: www.amjbot.org/cgi/content/abstract/88/11/2101.

Hill, M. O. 1973. Diversity and evenness: A unifying notation and its consequences. Ecology **54**: 427–431.

Hirao, A. S. and G. Kudo 2004. Landscape genetics of alpine-snowbed plants: Comparisons along geographic and snowmelt gradients. Heredity **93**: 290–298.

Hunter, M. L., Jr. 1990. Wildlife, forests, and forestry: Principles of managing forests for biological diversity. Prentice Hall, Englewood Cliffs, New Jersey.

Hunter, M. L. Jr. (ed.) 1999. Maintaining biodiversity in forest ecosystems. Cambridge University Press, Cambridge.

Hunter M. L. Jr., and J. P. Gibbs. 2007. Fundamentals of Conservation Biology. Blackwell, 3rd. edition. Malden, Massachusetts.

Jacobson, S. K. 1990. Graduate education in conservation biology. Conservation Biology **4**: 431–440.

Jacobson, S. K. 1999. Communication skills for conservation professionals. Island Press, Washington, DC.

Jarvis, A., M. E. Ferguson, D. E. Williams, et al. 2003. Biogeography of wild *Arachis*: Assessing conservation status and setting future priorities. Crop Science **43**: 1100–1108. Available online at: www.ciat.cgiar.org/biblioteca/pdf/ajarvis.pdf.

Jarvis, A., K. Williams, D. Williams, et al. 2005. Use of GIS for optimizing a collecting mission for a rare wild pepper (*Capsicum flexuosum* Sendtn.) in Paraguay. Genetic Resources and Crop Evolution **52**: 671–682.

Jules, E. S. 1998. Habitat fragmentation and demographic change for a common plant: trillium in old-growth forest. Ecology **79**: 1645–1656.

Karanth, K. U. and J. D. Nichols (eds.) 2002. Monitoring tigers and their prey: A manual for researchers, managers, and conservationists in tropical Asia. Centre for Wildlife Studies, Bangalore.

Karanth, K. U. and J. D. Nichols. 1998. Estimation of tiger densities in India using photographic captures and recaptures. Ecology **79**: 2852–2862.

Karanth, K. U., J. D. Nichols, N. S. Kumar, et al. 2004. Tigers and their prey: Predicting carnivore densities from prey abundance. Proceedings of the National Academy of Sciences **101**: 4854–4858.

Kingston, T., C. M. Francis, Z. Akbar, and T. H. Kunz 2003. Species richness in an insectivorous bat assemblage from Malaysia. Journal of Tropical Ecology **19**: 67–79.

Kisling, V. N. (ed.) 2001. Zoo and aquarium history. CRC Press, Boca Raton, Florida.

Krebs, C. J. 1999. Ecological methodology. Addison Wesley Longman, Menlo Park, California.

Laing, S. E., S. T. Buckland, R. W. Burn, et al. 2003. Dung and nest surveys: Estimating decay rates. Journal of Applied Ecology **40**: 1102–1111.

Laurance, W. F. and C. Gascon 1997. How to creatively fragment a landscape. Conservation Biology **11**: 577–579.

Lens, L., S. Van Dongen, and E. Matthysen 2002. Fluctuating asymmetry as an early warning system in the critically endangered Taita Thrush. Conservation Biology **16**: 479–487.

Lindenmayer, D. B. and J. Fischer 2006. Habitat fragmentation and landscape change. Island Press, Washington, DC.

Lindenmayer, D. B. and J.F. Franklin 2002. Conserving forest biodiversity. Island Press, Washington, DC.

Longino, J. T., J. Coddington, and R. K. Colwell 2002. The ant fauna of a tropical rain forest: estimating species richness three different ways. Ecology **83**: 689–702.

MacArthur, R. H. and E. O. Wilson 1967. The theory of island biogeography. Princeton University Press, Princeton, New Jersey.

MacKenzie, D. I. and J. A. Royle 2005. Designing occupancy studies: General advice and allocating survey effort. Ecological Applications **42**: 1105–1114.

MacKenzie, D. I., J. D. Nichols, G. B. Lachman, et al. 2002. Estimating site occupancy rates when detection probabilities are less than one. Ecology **83**: 2248–2255.

MacKenzie, D. I., J. D. Nichols, J. E. Hines, et al. 2003. Estimating site occupancy, colonization, and local extinction when a species is detected imperfectly. Ecology **84**: 2200–2207.

MacKenzie, D. I., J. D. Nichols, N. Sutton, et al. 2005. Improving inferences in population studies of rare species that are detected imperfectly. Ecology **86**: 1101–1113.

MacKenzie, D. I., J. D. Nichols, A. J. Royle, et al. 2006. Occupancy estimation and modeling: Inferring patterns and dynamics of species occurrence. Elsevier Publishing, San Diego.

Magurran, A. E. 2004. Measuring biological diversity, Blackwell, Oxford.

Manel, S., M. K. Schwartz, G. Luikart, and P. Taberlet 2003. Landscape genetics: Combining landscape ecology and population genetics. Trends in Ecology and Evolution **18**: 189–197.

Margules C. R. and R. L. Pressey 2000. Systematic conservation planning. Nature **405**: 243–253.

Martin, A. P. and S. R. Palumbi 1993. Body size, metabolic rate, generation time, and the molecular clock. Proceedings of the National Academy of Sciences of the United States of America **90**: 4087–4091.

McKee J. K., P. W. Sciulli, C. D. Fooce and T. A. Waite 2004. Forecasting global biodiversity threats associated with human population growth. Biological Conservation **115**: 161–164.

Morrogh-Bernard, H., S. Husson, S. E. Page, and J. O. Rieley 2003. Population status of the Bornean orang-utan (*Pongo pygmaeus*) in the Sebangau peat swamp forest, Central Kalimantan, Indonesia. Biological Conservation **110**: 141–152.

Myers, N., R. A. Mittermeier, C. G. Mittermeier, et al. 2000. Biodiversity hotspots for conservation priorities. Nature **403**: 853–858.

Newmark, W. D. 1987. A land-bridge island perspective on mammalian extinctions in western North American parks. Nature **325**: 430–432.

Norton, B. G., M. Hutchins, E. F. Stevens, and T. L. Maple (eds.) 1996. Ethics on the Ark. Smithsonian, Washington, DC.

Nunney, L. 1995. Measuring the ratio of effective population size to adult numbers using genetic and ecological data. Evolution **49**: 389–392.

Oertli, S, A. Muller, D. Steiner, et al. 2005. Cross taxon congruence of species diversity and community similarity among three insect taxa in a mosaic landscape. Biological Conservation **126**: 195–205.

O'Hara, R. B. 2005. Species richness estimators: How many species can dance on the head of a pin? Journal of Animal Ecology **74**: 375–386.

Orme, C. D. L., R. G. Davies, M. Burgess, et al., 2005. Global hotspots of species richness are not congruent with endemism or threat. Nature **436**: 1016–1019.

314

Otis, D. L., K. P. Burnham, G. C. White, and D. R. Anderson 1978. Statistical inference from capture data on closed animal populations. Wildlife Monographs **62**: 1–135.

Palmer, A. R. 1994. Fluctuating asymmetry analyses: A primer. In T. Markow (ed.), Developmental instability: Its origins and evolutionary implications, pp.335–364. Kluwer, Dordrecht. Available online at: www.biology.ualberta.ca/palmer/pubs/94Primer/94 Primer.htm.

Perlman, D. L. and G. Adelson 1997. Biodiversity: Exploring values and priorities in conservation. Blackwell Science, Cambridge, Massachusetts.

Possingham H., I. Ball, and S. Andelman 2000. Mathematical methods for identifying representative reserve networks. In S. Ferson and M. Burgman (eds.), Quantitative methods for conservation biology, pp. 291–305. Springer-Verlag, New York.

Primack, R. B. 2004a. A primer of conservation biology. Sinauer Associates, Sunderland, Massachusetts.

Primack, R. B. 2004b. Essentials of conservation biology. Sinauer Associates, Sunderland, Massachusetts.

Pullin, A. S. 2002. Conservation biology. Cambridge University Press, Cambridge.

Quammen, D. 1997. The song of the dodo: Island biogeography in an age of extinctions. Scribner, New York.

Reading, R. P. and S. R. Kellert 1993. Attitudes toward a proposed black footed ferret (*Mustela nigripes*) reintroduction. Conservation Biology **7**: 569–580.

Reid, W. V. 1998. Biodiversity hotspots. Trends in Ecology and Evolution **13**: 275–280.

Ries, L. and T. D. Sisk 2004. A predictive model of edge effects. Ecology **85**: 2917–2926.

Ries, L., R. J. Fletcher, Jr., J. Battin, and T. D. Sisk 2004. Ecological responses to habitat edges: Mechanisms, models, and variability explained. Annual Review of Ecology, Evolution, and Systematics **35**: 491–522.

Roberts, C. M., C. J. McClean, J. E. Veron, et al. 2002. Marine biodiversity hotspots and conservation priorities for tropical reefs. Science **295**: 1280–1285.

Root, T. L., D. Liverman, and C. Newman 2007. Managing biodiversity in the light of climate change: Current biological effects and future impacts. In D. W. MacDonald and K. Service (eds.), Key topics in conservation biology. Blackwell Science, Malden, Massachusetts.

Roslin, T. 2002. So near and yet so far – habitat fragmentation and bird movement. Trends in Ecology and Evolution **17**: 61.

Schleibe, S., T. Evans, K. Johnson, et al. 2006. Range-wide status review of the polar bear (*Ursus maritimus*). Unpublished report of the US Fish and Wildlife Service, Anchorage, Alaska. Available online at: http://alaska.fws.gov/fisheries/mmm/polarbear/pdf/Polar_Bear_%20Status_Assessment.pdf

Schweiger, O., M. Frenzel, and W. Durka 2004. Spatial genetic structure in a metapopulation of the land snail *Cepaea nemoralis* (Gastropoda: Helicidae). Molecular Ecology **13**: 3645–3655

Seber, G. A. F. 1982. The estimation of animal abundance and related parameters. Macmillan, New York.

Shrestha, R. K. and J. R. R. Alavalapati 2006. Linking conservation and development: An analysis of local people's attitude towards Koshi Tappu Wildlife Reserve, Nepal. Journal of Environment, Development and Sustainability **8**: 69–84.

Sih, A., B. G. Jonsson, and G. Luikart 2000. Habitat loss: Ecological, evolutionary and genetic consequences. Trends in Ecology and Evolution **15**: 132–134.

Simpson, E. H. 1949. Measurement of diversity. Nature **163**: 688.

Smith, D. M., B. C. Larson, M. J. Kelty, and P. M. S. Ashton 1997. The practice of silviculture: Applied forest ecology, 9th edn. John Wiley & Sons, New York.

Smith, J. N. M. and J. J. Hellmann 2002. Population persistence in fragmented landscapes. Trends in Ecology and Evolution **17**: 397–398.

Smith, R. J, R. D. J. Muir, M. J. Walpole, et al. 2003. Governance and the loss of biodiversity. Nature **426**: 67–70.

Soulé, M. 1985. What is conservation biology? BioScience **35**: 727–734.

Spear, S. F., C. R. Peterson, M. D. Matocq, and A. Storfer 2005. Landscape genetics of the blotched tiger salamander (*Ambystoma tigrinum melanostictum*). Molecular Ecology **14**: 2553–2564.

Stephens, S. E., D. N. Koons, J. J. Rotella, and D. W. Willey 2003. Effects of habitat fragmentation on avian nesting success: A review of the evidence at multiple spatial scales. Biological Conservation **115**: 101–110.

Stiassny, M. L. J. 1992. Phylogenetic analysis and the role of systematics in the biodiversity crisis. In N. Eldredge (ed.), Systematics, ecology, and the biodiversity crisis, pp.109–120. Columbia University Press, New York.

Stiassny, M. L. J. and M. de Pinna 1994. Basal taxa and the role of cladistic patterns in the evaluation of conservation priorities: A view from freshwater. In P. L. Forey, C. J. Humphries, and R. I. Vane-Wright (eds.), Systematics and conservation evaluation, pp. 235–249. Clarendon Press, Oxford.

Sutherland, W. J. 2000. The conservation handbook: Research, management and policy. Blackwell Science, Oxford.

Tallmon, D. A., E. S. Jules, N. J. Radke, and L. S. Mills 2003. Of mice and men and trillium: Cascading effects of forest fragmentation. Ecological Applications **13**: 1193–1203.

Thomas, L., S. T. Buckland, K. P. Burnham, et al. 2002. Distance sampling. In A. H. El-Shaarawi and W. W. Piegorsch (eds), Encyclopedia of environmetrics. John Wiley & Sons Ltd, Chichester.

Van Dyke, F. 2002. Conservation biology: Foundations, concepts, applications. McGraw-Hill, Boston, Massachusetts.

Vane-Wright, R. I., C. J. Humphries, and P. H. Williams 1991. What to protect? Systematics and the agony of choice. Biological Conservation **55**: 235–254.

Walsh, P. D. and L. J. T. White 2005. Evaluating the steady state assumption: simulations of gorilla nest decay. Ecological Applications **15**: 1342–1350.

Wieczorek, J., Q. Guo, and R. J. Hijmans 2004. The point-radius method for georeferencing point localities and calculating associated uncertainty. International Journal of Geographic Information Science **18**: 745–767.

Wilcove, D. S., D. Rothstein, J. Dubow, et al. 1998. Quantifying threats to imperiled species in the United States. BioScience **48**: 607–615.

Williams, B. K., J. D. Nichols, and M. J. Conroy 2002. Analysis and management of animal populations. Academic Press, San Diego.

Wright, S. 1931. Evolution in Mendelian populations. Genetics **16**: 97–159.

Wright, S. 1938. Size of population and breeding structure in relation to evolution. Science **87**: 430–431.

Zar J. H. 1999. Biostatistical Analysis. Prentice Hall, New Jersey.

Index

Index entries are arranged in word-by-word sequence; page numbers in italics refer to figures; page numbers in bold refer to tables.